Our Precious Metal

OUR PRECIOUS METAL

African Labour in South Africa's
Gold Industry, 1970–1990

WILMOT G. JAMES

David Philip: *Cape Town*
James Currey: *London*
Indiana University Press: *Bloomington*
& Indianapolis

First published 1992 in southern Africa by David Philip Publishers (Pty) Ltd, 208 Werdmuller Centre, Claremont 7700, South Africa

Published 1992 in the United Kingdom by James Currey, Ltd, 54b Thornhill Square, Islington, London N1 1BE, UK

Published 1992 in the United States of America by Indiana University Press, 601 North Morton Street, Bloomington 1N 47404, USA

ISBN 0-86486-165-6 (David Philip, paper)
ISBN 0-85255-217-3 (James Currey, paper)
ISBN 0-253-33092-0 (Indiana University Press, cased)

Printed and bound by Belmor Book Printers, Cape Town

Library of Congress Cataloging-in-Publication Data

James, Wilmot Godfrey, date

   Our precious metal : African labour in South Africa's gold industry, 1970-1990 / Wilmot G. James.
      p.   cm.
   Includes bibliographical references and index.
   ISBN 0-253-33092-0
   1. Gold miners–South Africa.   2. Gold industry–South Africa.
   I. Title
   HD8039.M732S646   1991
   331.7′6223422′0968–dc20

1 2 3 4 5 96 95 94 93 92

British Library Cataloguing in Publication Data

James, Wilmot G.
   Our precious metal : African labour in South Africa's gold industry, 1970-1990
   I. Title
   331.282234220968

   ISBN 0-85255-217-3

Contents

Preface

This book is a sociological history of African mine labour in modern South Africa. It is divided into two parts. The first part, *Organising a Labour Supply*, examines the labour recruiting strategies of the Chamber of Mines in the changing socio-political conditions of southern Africa. Changes in patterns of migrancy, such as the reduction of foreign labour, the increasing use of domestic labour, and the development of alternatives to migrant labour, are the empirical concerns of this part. The second part of the book, *The Ascendance of African Workers*, examines the processes by which African workers acquired power in the labour framework. Processes of unionisation, the struggles over mine compounds, and the erosion of the colour bar, are the empirical concerns of this part.

I was drawn to the subject matter while a visiting fellow at the Southern African Research Program at Yale University in 1985. At the time I was completing an article on the 1946 African miners' strike, which was essentially a story about how dominant groups in the mining industry and the state successfully defended labour-coercive institutions in the face of subordinate-group challenge. It struck me then that although we know a great deal about how labour-coercive institutions originated and became institutionalised, we know precious little about the conditions under which they might disappear. Stanley Greenberg's work, *Race and State in Capitalist Development*, was driven by similar concerns, but his study ended in the 1970s, before an organised African working class made its presence felt and directly challenged the labour framework. It made sense, therefore, to turn to the 1970s and 1980s when African workers developed class organisations, visibly asserted their interests, and remade the labour framework.

The research for this study was conducted in the larger Johannesburg area, where the head offices of the mining houses, trade unions, The Employment Bureau of Africa (TEBA) and the Chamber of Mines are located. A number of individuals in the corporate and trade union worlds were very helpful, and shared with me information and insights without which this study would not have been possible. Because of the current nature of the narrative contained here, and though I dearly wish to thank them publicly, their wishes to remain anonymous will be respected. There were also individuals who were very unhelpful, even obstructionist. Fortunately, their best efforts at suppressing information did not succeed in undermining the research.

The research was conducted between 1986 and 1989, when South Africa and its apartheid system were under considerable pressure from mobilised subordinate groups. A state of emergency, first proclaimed in 1985, was in place throughout the period. In itself, political repression did not interfere directly with the research, except in the case when the South African Police surrounded the now-derelict Congress of South African Trade Unions (COSATU) house in 1987, made access to members of the National Union of Mineworkers difficult, and destroyed some of their records. The repressive climate had paradoxical consequences. Some individuals in corporations, the state, and trade unions were extremely willing to talk to me, many driven by the desire simply to be better understood. Others refused, some because of racism (one corporate executive assumed that simply by virtue of my colour I would become a conduit of information to the National Union of Mineworkers), and others because they regrettably did not see much value in sociological research.

The Ford Foundation of New York generously funded all research costs, including three field trips to Johannesburg undertaken in 1986, 1987 and 1988. Without the Foundation's unfailing support, this study would not have appeared. The (University) Research Committee of the University of Cape Town provided supplementary funds for the final field trip. The Department of Sociology of the University of Cape Town tolerated my periodic absences.

Individual chapters were presented at the annual congress of the Association for Sociology in South Africa (Bellville, 1987), the annual congress of the Canadian Association for African Studies (Kingston, 1988), the seminar of the Centre for African Studies (University of Cape Town, 1988), the seminar of the African Studies Institute (University of the Witwatersrand, 1989), the Political Economy Workshop (Department of Sociology, Indiana University, 1990), the seminar of the African Studies Center (Indiana University, 1990), the seminar of the African Studies Unit (Queen's University, Canada, 1990), the Southern African Research Program Workshop (Wesleyan University, 1990) and the Council for African Studies (Yale University, 1990).

A number of colleagues read various drafts of chapters and I am very grateful for their comments: Jonathan Crush, Larry Griffin, David James, Gregory Hooks, Alan Jeeves, Jeffrey Lever, Dunbar Moodie, Michael Savage, Nigel Unwin, Charles van Onselen and David Yudelman. Alan Jeeves gave me a remarkable twenty pages of single-spaced commentary, and with the criticisms of an anonymous publisher's reviewer, compelled a complete redrafting of the manuscript. At the time I was overwhelmed by the breadth of revisions suggested by their criticisms, but now believe the study to be infinitely improved as a result of grappling with the complex issues they raised.

In Johannesburg I was based at the African Studies Institute, University of the Witwatersrand. The Institute's director, Charles van Onselen, went out of his way to provide access to facilities a visiting researcher requires. Friends in Johannesburg were extremely hospitable; Dee and Ewald Anderson, Simon Bekker, Adrienne Hall, Tessa Lever, and Dee and Nigel Unwin.

The manuscript was prepared on two continents. The first draft was completed in Paarl, South Africa. Located in one of Africa's most extraordinary settings, and deeply imbued with cultural contradiction, the Daljosaphat Art Foundation of Paarl provided a stimulating environment in which to write. I am indebted to the art of my wife, Julia Teale, whose talent was the reason for our presence at the Foundation, and for pushing me beyond the narrow creative boundaries of social science. I began the revisions of the manuscript while a visiting professor with the Department of Sociology and the African Studies Center at Indiana University, Bloomington, during their Spring semester of 1990. The revisions were completed and the manuscript finally prepared for publication in Cape Town.

Rose Kovats prepared the maps and Birga Thomas the manuscript for typesetting. Russell Martin, the publisher's editor, did a superb job of turning my social science prose into a much more readable and coherent product.

My father, Pat James, insisted on the value of academic work, and took an abiding interest in this research. He was always encouraging, and unfailingly supportive. He died in February 1988. It is to his life that this book is dedicated.

For Pat James

1
Our Precious Metal

The Mine Labour Framework

For over a hundred years, African migrant workers have come to the Witwatersrand to produce gold.[1] Out of the furthest corners of the sub-Saharan countryside, from the rural areas and reserves of South Africa, from what are now known as Lesotho, Botswana, Swaziland, Mozambique, Zimbabwe, Angola, Malawi and Tanzania, they have come in search of work and cash wages. Were it not for their labour, the gold mines of South Africa would not have become the world's most prolific producers of gold in modern history, nor would Johannesburg and the Witwatersrand stand as the industrial giant of the sub-continent.[2]

At first, would-be African miners were reluctant and erratic migrants. They had to be nudged out of their rural world, and prised away from their involvement in subsistence agriculture. Tax laws, imposed land-tenure arrangements, the expansion of markets, and growing indebtedness drove them in search of cash wages.[3] These were the 'push' factors, but they were not enough to sustain a reliable supply of workers. The Chamber of Mines, formed in 1889 to serve the collective interests of the Witwatersrand mine-owners, soon established centralised labour-recruiting agencies to search actively for labour. The Witwatersrand Native Labour Association (WENELA) was set up in 1897, with the objective of finding, initially, Mozambican workers and, later, workers elsewhere in Africa, while the Native Recruiting Corporation (NRC), formed in 1912, was geared towards finding domestic labour, mostly in the reserves of South Africa. The principal goals of these labour bureaucracies were to control supply and eliminate competition for labour between the mines.

By means of a standard contract, the recruiting agencies fixed the African workers' food supply, length of shifts, their working day, and wage rates. In terms of the so-called maximum-average system as it became known later, individual mines could settle their own wage scales, but these might not on average be higher than those of other mines.[4] Wages could in this way be fixed, free of the spiralling consequences of competition. The Chamber used this system to lower wages after the South African War of 1899–1902, and in the long run it ensured that real wages for miners would remain constant between 1911 and 1969.[5]

In his study of migrant labour, Alan Jeeves noted that the recruiting agencies had first of all to root out the small-scale but pervasive competition that labour touts provided. The touts brought African migrants to the mines more cheaply and efficiently than the Chamber's agencies.[6] Only by the 1920s did the Chamber finally succeed in monopolising the labour supply, after the labour touts had been driven out of the recruiting business.[7]

The emergent South African state provided a supportive political framework for the migrant labour system that was taking form. The 1913 Natives Land Act restricted African ownership of land to about 8 per cent of South Africa's surface territory, and laid the foundation for the overcrowded reserves which the Chamber came to defend, as early as 1920s, as the tribal 'homelands' of Africans.[8] Moreover, already in 1895 the Transvaal Volksraad had adopted a pass law drafted by the Chamber, whereby the movement of Africans from reserve to town could be policed by the state. Effective control of African movement only came, however, with the development of an efficient bureaucracy when the Transvaal administration was reorganised under Lord Milner after the South African War.[9] The Natives (Urban Areas) Act of 1923 extended the pass laws to South Africa as a whole, and further tightened restrictions on Africans entering the 'white' cities. Although it never acquired the status of law or written regulation, it became convention with state officials that no more than 3 per cent of the African labour force could live with their families on mine property. Other legislation, such as the Native Labour Regulation Act of 1911, gave the Chamber considerable leverage in the labour market, by criminalising African participation in work stoppages and strikes.

Able to fix African wages at set rates across the mines, the Chamber's recruiting agencies went into sub-Saharan Africa in search of men willing to work, from wherever they could be found. They hunted far and wide, building a colossal labour empire in the process.

However, while a relatively stable labour system developed over time, the Chamber was not able to rely permanently on any one supply area. Indeed, as David Yudelman and Alan Jeeves have observed, the instabilities of the supply pattern in the twentieth century often bordered unnervingly on crisis.[10]

In the 1920s and 1930s, for example, the mines employed significantly more South African than foreign workers. By the early 1940s, for a number of war-related reasons, the mines began to experience a shortage of labour, and the proportion of South African workers dropped off. At this time the Chamber initiated a three-decade-long programme to recruit more foreign labour, focusing its attention on workers particularly from Central Africa. By the 1960s and early 1970s, foreign workers dominated the mines, making up over 70 per cent of the labour force by 1970, the larger majority of whom were from Malawi and Mozambique.

Once at the mines, the migrant workers were compelled to live in compounds – barrack-like structures that housed up to ninety persons in rooms lacking even the most elementary forms of privacy. Adapted from the closed compound of the Kimberley diamond-fields, the open compound of the gold mine allowed workers to come and go under the supervision of management, which also controlled access by strangers.[11] In Kimberley the closed compound was primarily used to check diamond theft. On the Witwatersrand the open compound became a means of labour discipline. The compounds were located close to the mine shafts, making rapid and efficient mobilisation of labour for work possible. Furthermore, they ensured that mass feeding in compounds provided a fit work-force, absenteeism would easily be policed, and strikes could be broken with a minimum of effort.[12]

African migrant workers could neither establish nor belong to a trade union. The Industrial Conciliation Act of 1924 recognised unions for white workers, but it did not cater for African workers. While the legislation did not ban African unions outright, it had much the same effect by refusing them recognition if and when they were formed. The Chamber in any case had no time or enthusiasm for African trade unions, and actively suppressed African labour organisations and collective action. In 1920, for example, when African workers struck against bad food, low wages and the colour bar, the Chamber put the action down.[13] In 1946, again, the Chamber together with the South African Police suppressed a strike of 70 000 workers against inadequate food and low wages.[14] Furthermore, workers were routinely threatened with prosecution under the Native Labour

Regulation Act and often hauled before the magistrates for violating it.

In place of unions, the Chamber developed systems of worker representation organised along pseudo-tribal lines, within a paternalistic industrial-relations framework. In the Chamber's mind, African migrants were considered to be wards entrusted to the guardianship of mine management. They were not, the Chamber's ethnologists argued, mature enough for the advanced responsibilities of modern trade unions. Instead they had to make use of their *indunas* (or ethnic representatives) and *isibondas* (or room representatives) to advance their interests, and to convey their grievances to management. Only by the 1960s was there some movement away from tribally based to bureaucratically centred systems of representation, but these were still far from constituting independent trade-union representation.[15]

Furthermore, African workers could not advance beyond unskilled and marginal categories of semi-skilled work. A colour bar reserved the skilled, more senior and most renumerative jobs for whites. In terms of the Mines and Works Act of 1911, certain types of work required the incumbent to hold a certificate of competency, and these were awarded only to white workers. After a successful court challenge to the mine colour bar in the early 1920s, the Act was rewritten in 1926 to make doubly sure that only whites could hold the certificate, by building racially discriminatory clauses explicitly into the legislation.

A closed-shop agreement signed with the white unions in 1937 obliged all skilled workers, artisans and officials to belong to specified unions. An Allocation of Occupations Agreement defined in detail which jobs belonged to which unions. In turn, custom and convention added to what had become a ubiquitous and legalised system of job reservation. Moreover, it was general practice in the trades not to indenture African apprentices; the mines were no exception. Although the legislation governing apprenticeships was nominally colour-blind, the apprenticeship committees instituted their own race bar by refusing to indenture Africans.

In broad outline, this was the mine labour framework which was constructed over seven decades and which has survived into the modern period. It was a coercive and repressive labour system in that the rights of African workers regarding conditions of employment, housing, accommodation, collective organisation and trade unionism were circumscribed, even suppressed, by corporate management or the state and frequently by both.[16] On a day-to-day level, the various elements that made up the labour-repressive framework helped to generate and mould a militaristic and authoritarian industrial culture. Up until the 1970s, this was the world African miners encountered and came to know intimately.

Remaking the Labour Framework

In the early 1970s, there were several indications that this labour framework was about to change. Already in 1968, the system by which the international price of gold had remained fixed at US$36 since 1936 was partially abandoned, completely so by 1972. In 1974 Hastings Banda, the president of Malawi, withdrew over 80 000 workers from the mines, and in 1975 fears arose in Chamber circles that newly independent Mozambique might do the same. In 1977, the South African state appointed two labour commissions of inquiry (the Riekert and Wiehahn commissions) with the brief to review legislation governing employment discrimination, industrial relations and labour mobility in industry.[17] These events signalled the advent of structural changes in the political economy of gold-mining, in the socio-political conditions of southern Africa, and the shifting character of the South African state. Though apparently unconnected, they were to set in motion processes that profoundly affected and partially transformed the labour-repressive framework that had been constructed over the previous seven decades.

One of the most significant changes was the increasing shift towards domestic labour. This came about in the early 1970s when the Chamber faced its most serious labour crisis ever. Developments in Malawi and Mozambique, the two major external labour-supplying states, created great uncertainty in the migrant labour market. Members of the Chamber at the time believed that had Lesotho also withdrawn its labour, the gold mines would have come to a complete standstill. They felt that the industry had become too dependent on foreign workers, and decided to turn to domestic labour for their future needs. They also believed, however, that it was dangerous to rely on any single source, and that labour pools should be spread as widely as possible. Chamber members spoke of the need to create and maintain a desirable mix of workers, and not to become dependent on any one state or locality, foreign or domestic.[18] At the same time, the mines required their due quota of labour, and it was no foregone conclusion that the new sourcing strategies of the Chamber would meet the daily labour needs of individual mines.

Faced with major labour shortages, the Chamber signed an agreement in 1974 with the Rhodesian government to employ up to 20 000 jobless urban Africans every year. However, once it became clear that the labour supply from South Africa's homelands was expanding, the Chamber began to restrict the flow of workers from foreign states. Mozambique's Labour Ministry, confronted with a growing civil war and a faltering socialist economy, wanted to return to earlier, colonial patterns of migrancy, in terms of which close to 100 000 Mozambican workers were employed by the mines. Representatives of its Labour Ministry met annually with Chamber committees, where the request to send more labour was unfailingly presented in the early 1980s.

As it turned out, Malawi could not absorb all of its withdrawn workers in local employment, and by 1978 was pressing the mines to resume hiring about 20 000 workers annually. More secure in the migrant labour market, the labour ministries of Lesotho, Botswana and Swaziland nevertheless became concerned about the trend towards internalising the source of labour within South Africa, and closely monitored recruiting levels in their own countries. In the early 1980s the Chamber regularly received delegations from Lesotho, the largest foreign supplier in the 1980s, which expressed concern about the internalisation policies that the Chamber was adopting in labour sourcing. All the foreign states had for decades sent labour to the mines of South Africa, they had become dependent on the migrant-based revenues, and resisted the trend towards internalisation as best they could.

The Chamber's power to restrict the flow of foreign labour was rooted in its ability to create and exploit labour pools in South Africa's reserves. The remarkably improved gold price after 1973 changed the conditions of accumulation in the industry, making higher wages possible. In the 1970s, the Chamber and the mining houses increased African wages eleven-fold over a period of five years. At this time, too, as a result of the state's resettlement policies, the homelands had become densely populated areas where unemployment reached high levels. Rather than part-time peasants and target-workers,[19] homeland villages now contained a rural proletariat that was desperate for work, and the mines became overwhelmed by the number of willing workers from the homelands. As a result, the Chamber's recruiting agencies were able to pursue selective labour-hiring policies, and stabilise its labour force, becoming in the process gatekeepers and transporters rather than active recruiters of labour.

In line with these developments the Chamber became involved in the development of urban labour markets. In an earlier phase, recruiting offices had been opened on the Witwatersrand in 1974, but were closed in 1976 because of an apparent lack of worker interest. Rand Mines experimented with township workers at its ERPM mine in the early 1980s, but could not retain any number in mine employment. Later, Anglo American Corporation, Johannesburg Consolidated Investment (JCI) and Rand Mines expanded family housing and the employment of township-based workers in the mid-1980s. The violence in the compounds, the erosion of the colour bar and pressure from the National Union of Mineworkers (NUM) were some of the reasons behind these developments. Moreover, the state abandoned influx control in 1987 and also dropped the 3 per cent limit on family housing on the mines, thereby clearing the way for the possible expansion of family housing. However, other state restrictions on urbanisation (such as those pertaining to the availability and ownership of land) and the sheer cost to capital and state of settling workers on or near mine property, were to limit the expansion of these initiatives. By the early 1990s, only 5 per cent of the African labour force was involved in the emerging alternatives to migrant labour.

In the late 1970s, in another significant development that altered the nature of the migrant labour framework, the South African state began to withdraw from regulating industrial relations in the mining industry. The Durban strikes of 1973 and the emergence of an independent African trade-union movement spelt the end of the worker committees and liaison committees which had for long formed the basis of the official industrial-relations system. When it became apparent that attempts by the state to rescue and modernise its paternalistic labour apparatuses had failed, the Wiehahn Commission recommended that the state withdraw from industrial relations, and thereby allow companies to establish private systems of collective bargaining with African workers, albeit operating within a state-sanctioned industrial relations framework. What is more, Wiehahn recommended that African workers be permitted to establish independent trade unions, thereby reversing a century-old labour policy.

The Chamber was divided on the question of African unionisation. While its strongest member, Anglo American, was in favour of an open unionisation policy, with Anglo's junior partner, JCI, and Rand Mines agreeing, officials from General Mining Corporation (Gencor), Anglovaal and Gold Fields wanted to restrict unionisation to the non-migrant portion of the labour force. Divisions in the Chamber emerged as soon as it was faced with the task of constructing its own privately run industrial-relations apparatus. After considerable internal conflict, the Chamber reluctantly went along with Wiehahn's recommendations, hoping that African miners would not be drawn to trade unionism.[20] The National Union of Mineworkers took root among the work-force and developed rapidly in membership and strength.[21] Life in the mines would never quite be the same again. Wage-bargaining replaced wage-fixing, a bureaucratic system of industrial relations supplanted racial paternalism, and in general African workers became more assertive about their interests.

However, the growth of the NUM and the new assertion of worker interests were constrained by the flooded labour market. When workers went on strike in the 1980s, management's typical response was to dismiss them *en masse* and replace them with other, easily available labour.[22] During the 1987 African miners' strike, migrant workers crowded TEBA offices in the sending areas.[23] Though the mines were hurt by the strike, management had little trouble finding scabs and replacement labour, which was mainly provided by the homelands and foreign states. What disadvantaged the NUM was the fact that hired replacements were non-union members. Indeed, by the late 1980s, the union lost the recognition it had earlier gained in a number of job-groups and at a number of mines.

Paradoxically the new-found power of African mine workers was rooted in the mine compound. Although NUM leaders objected to the mine compound in principle (as part of their more general critique of migrant labour as exploitative and oppressive), in practice the mine-based leaders used the compound as a form of emergent worker-power. Once union organisers were granted access to the compounds, these changed from forming a strong barrier against successful organisation to becoming a facilitator of it. During the 1987 strike, mine-based strike and branch committees of the NUM used the compound to check the power of management and to discipline the labour force. Once the strike was over, however, management would regain control. Workers' control proved tenuous and limited, lasting merely for the duration of the strike. After the strike, managerial authority was intensified: union officials were denied access to the compounds, and meetings were forced off mine property. While debates emerged within some of the mining houses as to whether more cooperative and less authoritarian systems of compound management should be introduced, there was no doubt as to who ran the compounds.

The third important change to the migrant labour framework involved the abolition of the colour bar. One of the Wiehahn Commission's recommendations was that the state remove the colour bar in the mining industry. There was long-standing unanimity among members of corporate management that the colour bar protected indefensible and costly white privilege. For its part the NUM, representing an opportunity-starved and potentially mobile African work-force, participated in the political proceedings that led to the abolition of the colour bar. Both the Chamber and the NUM wanted the state to withdraw completely from the allocation of work in the industry. Labour allocation should be a privately negotiated matter, their representatives argued.

State officials, however, came under considerable pressure from white workers and their representatives to maintain a presence in the allocation of work, so as to safeguard the interests of white workers in the occupational structure. State officials felt restrained, too, by the increasingly visible participation of white workers in right-wing racist politics. The Minister of Mineral and Energy Affairs vacillated between various legislative proposals which at times protected and at other times abandoned white worker interests. In 1988 the colour bar legislation was finally repealed, but the Minister introduced regulations which, though non-discriminatory on paper, proved racially restrictive in practice. When the Chamber objected that these were tantamount to state intervention on behalf of white workers in non-racial guise, and challenged the Minister in the Supreme Court, the court struck down the regulations on the grounds that they exceeded the Minister's powers. In this way, the Chamber tried to limit state intervention in an area increasingly regarded as the sole and private preserve of capital and labour, under circumstances in which state officials still felt sensitive to the needs of white workers in the labour market.

By the late 1980s the labour framework of the gold mines had been transformed. A hallmark of the modern period has been the emergence of organised African labour and the assertion of its interests in the mine industry's labour framework. The unionisation of the labour force, the struggles over domestic life and managerial prerogatives in the compounds, and the erosion of the colour bar – the processes which this section has sought to introduce – increasingly made manifest the industrial presence of a work-force long hidden from view and obscured in history.

A Theoretical Framework

The development of organised African labour in the mining industry, and the limitations placed on the exercise of its power by a wider labour market and set of inter-state relations, define the theoretical interests explored in this book. How do we account for the processes by which workers in a labour-repressive framework acquire the capacity to change and modify their socio-political environment? What makes it possible for a migrant labour system to be reproduced despite the empowerment of African labour? How does the nature of the labour market and inter-state politics in southern Africa limit the power of organised African labour in the mines? What are the limits of social change in the mining industry's labour framework, and in South Africa more generally?

At the risk of some over-simplification, it can be said that the theoretical literature dealing with labour-repressive institutions in southern Africa tends to fall into one of three traditions. The first, and historically the most influential, approach focuses on reproductive processes in capitalist societies – a reproduction-centred analysis of labour. In this approach the development and persistence of labour-repressive institutions are considered in terms of their functional consequences for the migrant labour system, for patterns of accumulation, and capitalist development and growth.

There are two versions of this approach. Harold Wolpe, in his well-known essay, argued that the effects of labour-repressive institutions were to cheapen African labour, and to create and maintain a low-wage labour system.[24] According to his analysis, changes in the labour framework could be understood in terms of the changing requirements needed to reproduce low-wage African workers. Migrant labour, mine compounds, the lack of trade-union representation and the colour bar were all instrumental in reproducing the cheap labour system. Wolpe's analysis was valuable both because of the novel and controversial approach he brought to historical material, and because of the voluminous literature, both supportive and critical, generated by his cheap-labour analysis of apartheid institutions.[25]

A second version of the reproduction-centred approach is found in Michael Burawoy's earlier work on Zambian, South African and Californian migrant labour.[26] Burawoy did not dispute the low-wage character of migrant labour so much as argue that Wolpe's analysis was too narrowly focused on cheapness.[27] He suggested that there was a great deal more to be learnt about labour-repressive institutions if migrant labour was defined in institutional terms and not narrowly economic ones. Using a range of comparative material, Burawoy tried to show how an institutionally based conceptualisation of migrant labour allowed for a more complex and less economistic analysis.

Reproduction-centred analysis can generate useful questions about why social institutions persist. There are, however, a number of well-known problems associated with its uncritical application. Because of its emphasis on reproduction and function, there is a tendency to treat mining capital and the state in an undifferentiated and functionalist manner, and to underplay the divisions between dominant groups.[28] My study will indicate the critical importance of understanding the fissures within state institutions and the conflict between the mining groups in the 1970s and 1980s.

A framework too narrowly concerned with function and reproduction easily slips into a conceptualisation of social change that is attentive to institutions but indifferent to actors. In this view social change is seen to result from the loss of reproductive properties in institutions rather than the actions or behaviour of individuals and groups. Reproduction analysis can deny individuals and groups an active role in historical processes. It has great difficulty accounting for how subordinate groups attain power, assert interests, and remake social institutions. As a result, when applied to South Africa it pays no attention to African workers in the labour framework: instead they are reduced to the silent victims of reproductive processes.

It was in response to some of the shortcomings of reproduction analysis that a more state-centred literature emerged. Here the concern lay with the active involvement of dominant groups and state officials in the construction of labour-repressive institutions. A notable contribution along these lines was Stanley Greenberg's *Race and State in Capitalist Development*, which explored the relationship between labour-repressive state institutions, racial hierarchies and capitalist development in four settings: South Africa, Israel, Northern Ireland and Alabama, USA.[29] Greenberg argued that in these societies state officials elaborated and intensified labour-repressive institutions to assist dominant groups in procuring labour during early phases of capitalist development. Once dominant groups came to rely more directly on market processes to hire and retain labour, labour repression became less essential for accumulation purposes. The erosion of labour-repressive institutions is in this view, therefore, directly linked to the ascendance of markets and market processes.

Greenberg recognised that although he placed the state in the centre of his narrative, he nevertheless rendered it as a relatively undifferentiated and functional entity. In a later work he addressed this problem directly, and fleshed out the internal workings of state bureaucracies involved in the implementation of influx control.[30] Other state-centred studies, such as David Yudelman's historical work on the state and the gold mines, have similarly pointed to the centrality of state institutions and state officials in the construction (and erosion) of labour-repressive institutions.[31] Like Greenberg, Yudelman stressed the importance of understanding conflicts within the state, and the struggle of state officials to implement policy.

The state-centred literature is also sensitive to divisions and conflicts within dominant groups. In *Race and State*, Greenberg divided his account of dominant-class participation in constructing labour-repressive institutions along sectoral lines. Yudelman's narrative, in turn, emphasised lines of unity and division in the mining house structure, and showed how these dynamics influenced the policies of the Chamber of Mines. Other work, such as Merle Lipton's *Capitalism and Apartheid*, also looked to divisions within dominant classes to explain the erosion of labour-repressive institutions.[32]

The state-centred literature has a number of limitations, however. Divisions within the state are foregrounded, but left unexplored. Like the reproduction-centred analysis, the work on the state pays little attention to African workers. For example, in *Race and State* Greenberg focused on subordinate African groups only when they intrude on the public stage, such as with a major strike and work stoppage, or as part of visible episodes of political mobilisation, such as Sharpeville in 1960 and Soweto in 1976. Yudelman ignored African groups altogether in the main body of his narrative, presumably because they have no public presence in the state. An earlier, more marxist literature on the state did much the same, by equating public visibility with political importance, and by assuming that African victims of an oppressive order make no impact on social institutions.[33] In *Legitimating the Illegitimate*, Greenberg, while considering the impact of African workers on the labour market and the state's labour bureaucracies, left the emergence and development of class capacities among African workers largely undeveloped.

This lacuna would be dealt with by the labour studies of the 1980s, which focused attention specifically on the emergent class power of African workers. By synthesising labour history and industrial sociology, this approach sought to make sense of the emergence of the independent African trade-union movement in the 1970s and 1980s, the growth of worker power, and the organisation of production and labour processes. Though unevenly developed, this literature sought to document the rise of African workers in the labour framework in the 1980s.[34] A number of studies, for example, examined pre-existing formal and informal social networks of the larger mining world that facilitated unionisation processes.[35] Some initial studies of unionisation have appeared;[36] and the construction of a new industrial-relations system and the impact of class struggles on the labour framework have also been the subject of a number of works.[37]

On the basis of this research, we can now begin to explain how African workers developed class capacities in a labour-repressive framework. However, as Jon Lewis pointed out in a valuable review, the labour studies literature operated within a narrow frame of reference, rarely going beyond the workplace.[38] The impact of wider labour markets, state politics and inter-state relations on emergent worker power was left unexamined. In recognition of this, some studies have recently tried to make explicit links between struggles in the workplace and larger market and political processes. Rob Lambert, for example, has examined the relationship between workplace struggles, nationalism and proto-socialism among African workers.[39] Eddie Webster has spoken about 'social-movement' unionism to explain the largely political nature of the African trade-union movement. And in his comparative work, Michael Burawoy has used the concepts 'production politics' and 'global politics' to get at the complex relationships between the labour process and the state, struggles over work and struggles over state power.[40] These studies have all sought to bring the state back into production systems.

My own study underwrites the attempt to link the development of worker power with state power. But it is critical at the same time of the tendency to confine state power simply to the nation-state. Burawoy, for example, was dismissive of the impact of sub-continental and regional political relations on the migrant labour system.[41] His notion of 'global politics' ended with the nation-state, and failed to account for inter-state relations. My study will show how post-colonial states, capitalist and socialist, in southern Africa developed over time an interest in the sale of labour-power to South Africa, and reproduced migrant labour markets in the process.

The Sociologist as Historian

In his book *The Comparative Method*, Charles Ragin argues that there is much to commend the case-study approach to the study of social institutions.[42] Compared with other methods, he claims, the case-study approach lends itself to in-depth historical exploration, the interpretation of the specific meaning of things, and the explanation of why certain social institutions emerge and others do not.[43] Although a case-study deals with one particular society or set of institutions, it nevertheless can make general theoretical statements about social phenomena: it does not have to be parochial or celebrate merely the peculiar.

This book tries to make a contribution along these lines, inasmuch as its preoccupation with the detail of labour-repressive institutions in South Africa is also a concern with the reproduction and erosion of labour-repressive institutions more generally. However, what Ragin fails to mention is that in certain social settings, where written history of social institutions is poorly or unevenly developed, the sociologist using a case-study approach is often compelled to use also the methods and techniques of the historian.

South African mine-labour historiography is best developed on the earlier periods, such as the formative one of 1890 to 1920.[44] There is also considerable work on critical events, those that apparently mark the highpoints of historical processes, such as the 1922 white miners' strike and the 1920 and 1946 African miners' strikes.[45] Although a great deal on the modern period has been written on specialised topics, such as labour-recruiting patterns, the rise of African trade unions, mine safety, and the colour bar,[46] little of a comprehensive nature has been written on larger processes in the 1970s and 1980s.

General works tend to end before the changes in the labour framework began to take effect. Francis Wilson's classic book concludes in the 1960s, while Stanley Greenberg's material ends just when the labour order begins to 'unravel'.[47] Though David Yudelman's study contains an epilogue on the modern period, where he makes suggestive comparative comments about the rise of African workers in the labour framework, his book is essentially about early twentieth-century white labour history and the nature of the state.[48] Merle Lipton's *Capitalism and Apartheid*, though published in 1985, is based on research ending in the late 1970s and largely silent on the transformations of the labour framework after that.[49] Steven Friedman's work on the development of African trade unionism in the 1970s and 1980s is an extremely useful source, but by virtue of his focus has little to say about labour-repressive institutions in the gold industry, such as the mine compounds or the colour bar.

This study relies on original sources as the means of evidence, and more generally on historical methods of data collection. In addition to published sources, such as newspapers, company journals and annual reports, unpublished reports, memoranda, circulars, correspondence, statements of evidence and transcripts of meetings have been used to establish a data-base of information. In a society where the practising sociologist does not have access to a well-developed, historically oriented literature, he or she is compelled by circumstance to be an historian as well. Having gone this route, this work naturally falls into an emergent tradition in the social sciences known as historical sociology, where original source material and historical methods of data collection are used to answer sociologically motivated questions.[50] As I hope to show, the methodology of historical sociology is a compelling direction to pursue, given its dual emphasis on historical reconstruction and explaining more general social processes.

The Research Context:
The Gold Mines
in the 1970s & 1980s

A Political Economy of Gold

The gold mines of South Africa are located along a reef that begins in the eastern Transvaal, passes through Johannesburg to the western Transvaal, and curves all the way down to the Orange Free State. Six goldfields are situated along this reef – the East Rand, Central Rand, West Rand, Far West Rand, Klerksdorp and the Orange Free State goldfields (see Map 1). In the 1980s there were 43 mines producing gold, their presence in the landscape revealed by the conspicuous headgear supporting the elevator systems of the typical deep underground mine.

Between them, these six goldfields have over time made the single largest contribution to the world's gold production. In the 1980s, a third of the world's gold came from South Africa – an average of 600 out of 1 670 tonnes per year. The Soviet Union produced an estimated 300 tonnes, and Brazil, Canada, Australia and the USA some 80–160 tonnes each annually.[1] Compared to earlier periods, the quantity of gold coming out of South Africa in the last two decades has declined as a result of the mining of poorer ore-grades, a trend encouraged by the rising gold price which made their extraction profitable.

Gold has occupied a central place in the development of South Africa's political economy. Historically, accumulation patterns based on gold-mining have formed the engine of the economy's development, and a major source of state revenue. As David Yudelman indicated in his study of gold-mining in the early twentieth century, a mutually dependent 'symbiosis' developed between the gold mines and the state, and much of South Africa's economic history – including aspects of apartheid – has concerned the unfolding of this relation-

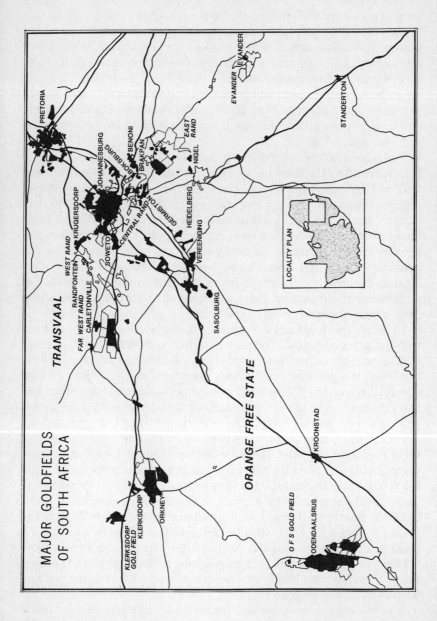

MAJOR GOLDFIELDS
OF SOUTH AFRICA

TRANSVAAL

WEST RAND
RANDFONTEIN
FAR WEST RAND
CARLETONVILLE

KRUGERSDORP
RANDFONTEIN
SOWETO
JOHANNESBURG
CENTRAL RAND
GERMISTON
BOKSBURG
BENONI
BRAKPAN
EAST RAND
NIGEL
HEIDELBERG

PRETORIA

VEREENIGING

SASOLBURG

ORANGE FREE STATE

KLERKSDORP
GOLD FIELD
KLERKSDORP
ORKNEY

O F S GOLD FIELD
ODENDAALSRUS

KROONSTAD

STANDERTON

EVANDER
EVANDER
EVANDER

LOCALITY PLAN

ship.[2]

Although the character of the South African economy has changed during this century in a number of important respects, the dependency of economy and state on the gold mines has persisted right into the 1990s. Gold has continued to provide the bulk of the country's foreign exchange, and a third of all export earnings. In the 1980s approximately 10 per cent of the nation's gross domestic product and 25 per cent of direct state revenues have been gold-based. Next to agriculture, gold-mining has been the single most important source of jobs in the economy. Over 700 000 individuals, mostly black, were employed by the gold mines, and over one million are in the mining industry as a whole.

The labour policies of the gold mines have historically been strongly influenced by the fixed price of gold, which held between 1935 and 1972 at US$36 per fine ounce. The fixed price meant that the mine-owners could not pass on any increase in production costs to consumers, but had to absorb it internally.[3] In the context of the country's racial order, where African workers were politically powerless and economically vulnerable, austerity measures were most easily targeted at their wage rates. As reproduction-centred theories have argued, coercive control over African migrant workers can be seen as an accumulation strategy, the cheapening of African labour being the mine-owners' answer to the fixed price.

In 1968, the fixed-price regime was replaced world-wide by a two-tier pricing system. The new system retained the fixed price (of US$36 per fine ounce) for central banks and monetary institutions, but freed the price for speculators, hoarders, and jewellery and industrial buyers. In 1972 the fixed price was abandoned altogether, and the price of a fine ounce was allowed to rise to a level regulated principally by market forces. The reasons for these changes, and the events surrounding them, have been described in Timothy Green's lucid narrative.[4] In brief, the financial institutions based in London were unable to maintain a gold pool substantial enough to underwrite the fixed price; instead the bankers introduced a two-tier system to maintain the market. As it happened, South Africa decided at about the same time to shift from London and market its gold in Zurich. The loss of their largest single supplier forced the bankers to abandon the fixed price altogether in 1972. As a result, the price of gold rose from US$36 per fine ounce in 1971 to reach over $800 by 1980. It settled down thereafter to a band fluctuating between $500 at its highest and $300 at its lowest (Figure 1).

The market price regime and the increases in the gold price made

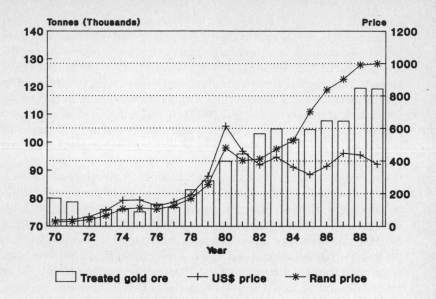

Fig. 1. Gold price and ore output, 1970–1989

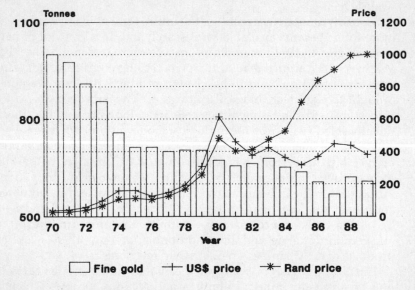

Fig. 2. Gold price and gold output, 1970–1989

a number of things possible. Firstly, mine-owners could now afford to pay much higher wages to their African migrant workers. As David Yudelman pointed out, the higher gold price was a necessary condition for the increases to take place.[5] What is more, the labour crisis and conflicts in the compounds over wage levels also contributed to the rise in wage rates. Between 1972 and 1980 there was an eight-fold increase, and between 1980 and 1987 a further three-fold increase (see Figure 3). In real terms, African wages improved by 320 per cent between 1972 and 1980, and by about 60 per cent between 1980 and 1987 (see Figure 4).

The wage-base of African workers prior to the increases had been extremely low. In 1970, African workers earned R16 per month, while on average white workers earned R337 per month. The relatively high increases in African wages and much lower increases for white workers narrowed but did not eliminate the racial wage ratio. In 1970 the white–African ratio stood at 22:1. By 1980 differential increases reduced it to 6:1. During the 1980s it dropped further to 5:1. After wage-bargaining for African workers was first introduced in 1983, real wages improved only marginally, however. It is ironic that when African workers became organised, the labour market turned against them, the over-supply of labour keeping a downward pressure on wage levels.

Secondly, as a further result of the improvement in the gold price, more funds became available for research and development of new technologies. In 1974 a memorandum was circulated in the Chamber of Mines which argued that the use of increasing numbers of domestic workers should go hand in hand with productivity improvements brought about by technological innovation.[6] The improved gold price at least made the investment possible, in an industry that had traditionally shied away from technological modernisation. Between 1974 and 1986, the Chamber of Mines Research Organisation had its annual budget increased from R4,66 million to R46,7 million, a ten-fold increase over the period.[7] Many of the technological innovations introduced in the industry in the 1980s came about as a result of the additional funding of research and development. The hand-held hydraulic drill, improved stope-support systems, underground refrigeration, cooling and air-conditioning, among others, were all products of the Chamber's new drive to mechanise.

Thirdly, the higher price of gold meant that poorer-grade ore could now be profitably mined, and mines and sections of mines which in the past could not be worked economically became a viable proposition. As the price of gold moved upwards, more low-grade ore was

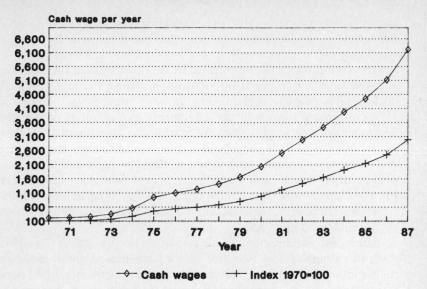

Source: Hirschson (1988)

Fig. 3 Annual African wages, 1970–1987 (average rands)

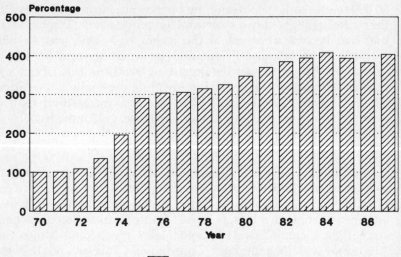

Source: Hirschson (1980)

Fig. 4 Annual African real wages, 1970–1987 (index 1970 = 100)

mined (see Figure 1). Because the mined ore was of declining quality, though, the actual quantities of gold produced decreased over time (see Figure 2).

One consequence of the mining of the lower-grade ore was the expansion of employment opportunities during the late 1970s, as additional workers were required to mine the new areas. In 1975 and 1976 employment levels increased sharply as a result as well of the Chamber's efforts to find as many workers as possible in the face of the labour crisis. Once the labour market settled down, there was a slow increase in annual employment in the later 1970s and 1980s, due principally to the declining grade of ore being mined. Almost all of the additional workers were hired from South Africa, in line with the Chamber's labour internalisation policies that have already been mentioned (see Figure 5).

After 1987 employment levels began to decline again. The 1987 African miners' strike was the immediate reason for this. Mine management fired about 50 000 workers, and selectively rehired non-union members; other workers had their employment-guarantee certificates withdrawn. Moreover, management succeeded in maintaining production levels with smaller work-forces in a number of telling instances. In all, between 1987 and 1990, the industry hired 60 000 fewer workers in its mining operations. Hiring practices after the strike highlighted two structural tendencies in the labour system that had become apparent in the recent past. One was the shift towards more labour-efficient and productivity-centred mining operations. The other was the persistent trend towards labour substitution through technological innovation, especially after African labour became organised. Both processes have increasingly reduced the number of jobs available to workers in the gold industry.

The Mining Houses and the Chamber of Mines

Although the 43 gold mines on the Witwatersrand are registered as individual companies, they are administered by six finance corporations or mining houses. Anglo American Corporation (AAC) administers thirteen mines, General Mining Corporation (Gencor or, since 1989, Genmin) twelve, Gold Fields nine, Rand Mines five, Anglovaal and Johannesburg Consolidated Investment (JCI) two each. JCI and Anglo American are linked corporations, the former effectively run as a junior partner of the latter. Of the finance houses Anglo American is the largest, and employs about 40 per cent of the mine labour force.

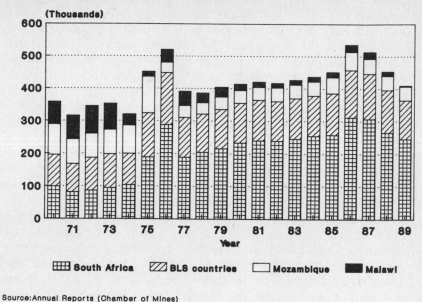

Source: Annual Reports (Chamber of Mines)

Fig. 5   Employment of African workers on gold mines, 1970–1989

The finance houses are not only involved in gold-mining, although that constitutes their major investment. They also administer coal, diamond, platinum, antimony, asbestos and copper mines; and uranium is mined as a by-product of some gold operations. The finance houses have also over the decades expanded into the industrial, banking, service and property sectors.[8] Their involvement in these sectors has made a significant impact on their corporate philosophies and policies, and, in the case also of Anglo American's foreign mining operations, explains in part the distinctive corporate attitudes towards migrant labour and the emergent African trade unions which the company expressed in the 1980s. Indeed, the fact that some of the mining houses had experience with African trade unions in their non-mining and foreign operations made it easier for them to come to terms with organised African labour in mining itself.

Representing the collective interests of mine employers is the South African Chamber of Mines, to which all the major gold mines are affiliated. The Chamber performs a number of functions for individual mines. All wage negotiations and collective bargaining with organised labour are managed centrally by the Chamber. Its employment agency, The Employment Bureau of Africa (TEBA), has recruited African migrant labour both for mines affiliated to the Chamber and for others not. The Chamber runs a research and development agency – the Chamber of Mines Research Organisation (COMRO) – and the industry's training schools, and administers three specialist-referred hospitals. The general affairs of the gold mines – marketing and labour, above all – are managed by the Chamber's Gold Producers Committee (GPC), consisting of representatives from each of the mining houses.

The co-operative structure of the mining houses and the centralised labour functions of the Chamber are both products of the gold industry's special history. The exorbitant expense required to develop a gold mine (an average cost in the 1980s of one billion rands) and the enormous risks of failure involved are the major reasons why the finance houses emerged. Through this system, large quantities of capital could be raised, and the risks cushioned by the extensive spread of investment the finance houses held.[9] At the same time, the mines' overriding need to hire cheap African labour brought it about that labour functions came to be vested in the Chamber. Centralised recruiting and standardised wage levels eliminated competition between the finance houses and between individual mines for labour, and as a result kept a downward pressure on earnings.

In the 1970s, the unity of purpose for which the Chamber had been renowned began to show signs of strain, and in the 1980s it broke down in a number of important respects. As will be shown later, division and conflict within the Chamber were important preconditions for the growing ascendance of African workers in the labour framework. Conceptually, when it came to African workers the Chamber's functions divided along two lines, namely (1) the organisation of the migrant labour supply outside the place of work, and (2) the management of a paternalistic system of industrial relations at the place of work. While differences existed between the mining houses regarding the composition by region of their African labour force, there was little conflict about the need to maintain the migrant labour system and associated recruiting bureaucracies. For this reason, the Chamber had no difficulty in maintaining its cohesion and consensus when it came to formulating policies regarding the organisation of the industry's labour supplies.

However, division and conflict between the mining houses were to emerge on the industrial relations side of the Chamber's functions. In the late 1970s, the state abandoned its paternalistic system of liaison and works committees, because of their political effects on production relations and capital accumulation, and their manifest inability to resolve ordinary and petty grievances. The state-appointed Wiehahn Commission recommended that African workers be granted the right to form and join unions, and that a private system of collective bargaining replace the traditional paternalistic one. When the various mining houses were confronted with the task of constructing a different system, there was strong disagreement as to what form it should take. Some of the mining houses wanted to extend trade-union rights to all African workers, while others sought to reserve the newly acquired rights for the proletarianised section – and therefore a minority – of the labour force. The division paralysed the Chamber, undermining its ability to put in place a new framework.[10] Indeed, Anglo's labour intellectuals at one point were willing to bypass the Chamber altogether when negotiations with the state over the new system were proceeding. The division was finally resolved after African workers had irrevocably established their trade unions in manufacturing and service industries, and when the state conceded full collective-bargaining rights to all African workers in response.

Once the National Union of Mineworkers came into being, other lines of division emerged in respect of wage-bargaining. After the maximum-average system broke down, the mining houses expounded increasingly diverse wage philosophies and introduced non-standardised wage scales. Although the Chamber still set average wage levels for African workers, the minimum and average wages of the mining houses began to differ. Consequently, the Chamber could not always obtain a consensus about wage increases for African workers. In 1985, for example, Anglo American for the first time broke ranks with the Chamber, and made a separate wage offer to the NUM. The union thereafter called its members out on strike at Gencor and Gold Fields mines. Subsequent wage-bargaining was to cause enormous strain, and the Chamber had to struggle to maintain a united front in the face of the NUM's challenge.

Thus, as we have seen, a close relationship exists between the Chamber and mining houses, built up over almost a century. In recent years, however, this relationship has begun to change, as individual mining groups were forced to respond to the breakdown in the traditional industrial relations system.

The Mining Experience

The gold-bearing ore of the Witwatersrand is found in a number of different reef-leaders that lie at an angle of about 23 degrees to the earth's surface. The thickness of the reef ranges from four inches to eight feet, averaging about twelve inches. The deposits are unevenly layered, shifting in their angular distribution, and are in places disjointed and broken. The job of the miner is get the gold-bearing ore out as efficiently as possible, bringing to the surface the minimum amount of waste-rock together with the actual gold. Given the geological nature of the deposit, this is by no means an easy or straightforward job.

The task of the gold-miner is perhaps best conveyed by the dictionary metaphor the Chamber of Mines uses in its public affairs publications: 'Imagine a solid mass of rock tilted ... like a fat 1 200 page dictionary lying at an angle. The gold bearing reef would be thinner than a single page, and the amount of gold therein would hardly cover a couple of commas in the entire book. It is the miner's job to bring out that single page – but his job is made harder because the "page" has been twisted and torn by nature's forces, and pieces of it may have been thrust between other leaves of the book.'[11]

A complicating factor is the increasing depth at which mining operations take place. The average working area is about one and a half miles (2 400 metres) underground, and the deepest mine, Western Deep Levels of Anglo American Corporation, reaches two miles (3 600 metres). The deeper mining goes, the hotter and more dangerous it becomes, and the more frequently workers are exposed to hazardous and disagreeable conditions of work.

The route miners have to travel to the working area is long and complex. The deepest mines use a staggered system of elevators to transport their workers underground, and two to three systems are often needed. Once the required depth has been reached, the miners travel an equally astonishing horizontal distance underground to reach the working face. The whole journey can take up to three hours, involving travel by elevator, train and foot through a labyrinth of shafts, cross-cuts and tunnels. A journalist who recently went underground described his surprise at the distances involved when visiting the Vaal Reefs mine in the western Transvaal: 'To get to the stope-face where the seam of gold runs, we hurtled down three separate shafts in cage-like lift contraptions, walked a couple of kilometres through winding tunnels, rode in a locomotive for another two kilometres and finally crouched and crawled to the stope-face.'[12]

According to work routine, the first shift of work begins early in the morning. Miners are expected to be at the shaft around 5 a.m., and because of the travelling distance involved they rise at about 3 a.m. African miners on shift are called to work over a public address system, are fed a breakfast, and then proceed to the shaft, from where they descend to the working face.

The area whence the gold-bearing ore is removed – the stope-face – is very noisy and hot, even though many measures have been taken to reduce noise levels and to cool the environment.[13] Crouched in an area no higher than about five feet and with very little headroom, miners drill the holes in preparation for blasting. Once a particular area has been blasted, the miners return to the workplace to clear the gold-bearing ore. The combination of heat, sweat and noise can be overwhelming. Francis Wilson wrote: 'it is perhaps easiest to start by thinking of a road labourer digging up a pavement with a jackhammer drill. Now imagine him doing that work thousands of feet underground, in intense heat, where he cannot even begin to stand upright.... Add to this picture the noise of a road-drill, magnified several times by the confined space; dust which, despite strenuous efforts to control it with water, invades the lungs.'[14]

This underground sensibility is pervaded by constant anxiety, and by fear of the unknown and unexpected. Rockfalls and rockbursts are common, fires and explosions can occur, and their frequency increases the deeper underground mining goes. There have been dreadful accidents, as in the case of the Kinross disaster in September 1986, when 177 miners lost their lives as a result of an underground fire. Other, smaller-scale accidents occur year after year. Although there has been some decline in deaths due to accidents underground, the cumulative mortality figures between 1905 and 1989 of 66 000 miners dead as a result of work-related accidents remains a chilling statistic indeed.[15]

Velile Maqakula, a African miner interviewed by a local newspaper, recalled that after experiencing a rockfall, he was 'not looking forward to going back underground'. He noted that 'even before the accident [he] was very tense and found it impossible to relax underground'.[16] This is not the stuff for the faint-hearted; the author, who is a city-based sociologist, came away from an obligatory visit to a working face 2 000 metres underground, convinced that no human being ought to work there, and impressed that a half a million people nevertheless do. People of the city, officials in the state, and a nation that depends on the wealth generated here, simply have no conception of what underground mining means to those who daily descend the shafts.

Although the basic features of the labour process in gold-mining have remained much the same over time, a number of technological changes introduced in the 1970s and 1980s have affected the ex-

perience of underground work for individual miners. A few mines
have introduced underground cooling and air-conditioning systems,
and have dispensed with the much-disliked acclimatisation exercises.
A new hand-held and much quieter hydaulic drill was first introduced in the late 1970s, and in the 1980s was widely in use. Again, in
a few mines and some mine shafts, trackless mining equipment is
being used to transport the ore from the working face to the elevators,
using quieter and safer rubber-tyred front-loading dump trucks instead of the less efficient and more dangerous train system.[17]

Other innovations have added a less harsh and safer working
environment, such as the use of backfilling (the refill of already
worked areas with slime), hydraulic underground roof-support systems, and water-powered equipment.[18] Generally, though technological changes have been unevenly introduced across the various
mines and shafts, and the overall experience of working conditions
underground varies greatly as a result, the conditions of work are less
severe compared to even two decades ago.

It must be remembered that the new technologies were not introduced simply to improve underground working conditions, but to
improve labour productivity, the two goals often, but not always,
going hand in hand. Trackless mining is said to enhance productivity
considerably; the hand-held hydraulic drill to drill much more efficiently than the older one. The introduction of these two technologies
has also resulted, however, in the retrenchment of labour. In particular, significantly fewer workers are required to run the trackless
mining equipment underground. JCI, which widely uses trackless
mining, cut its workforce of 40 000 down to 23 000 in the late 1980s;
its new mine, H. J. Joel will only have a labour force 2 500 strong.[19]
Thus new technologies have mixed consequences, on the one hand
reducing the size of the labour force but also improving underground
conditions of work and the mining experience for thousands of
workers.

Stability of African Labour

In the 1980s the Chamber succeeded in stabilising the migrant
labour system. 'Stabilisation' refers to regular employment patterns
in the migrant labour system, and should be distinguished from
'proletarianisation', a process where migrant workers begin to sever
their social ties with the rural areas.[20] Already in the past, as Jonathan
Crush has pointed out in his review of 'stabilisation' patterns, foreign
worker-groups tended to develop fairly stable employment patterns.

Their vulnerability in the labour market and stiff competition for jobs encouraged high levels of regular employment.[21] A new departure was marked in the aftermath of the foreign labour crisis of the 1970s when TEBA offered domestic workers shorter contracts, with a view to attracting them to mine employment. Once it became clear that an adequate supply of labour could be generated in the homelands, the recruiting agencies began to encourage stability among domestic workers, alongside the already established pattern prevailing among foreign workers. In the midst of a flooded labour market, TEBA wanted to create and maintain constant pools of workers employed from one contract to another.

The administrative mechanism used to encourage stabilisation is a revised, rationalised version of the older Valid Re-engagement Guarantee (VRG) system. This system had encouraged workers to remain at work for longer periods and at home for shorter. In the late 1970s, however, the Chamber introduced four standard industry certificates: TEBA 461 guaranteed re-employment and an early-return bonus, TEBA 460 guaranteed re-employment to workers with six to nine months' service, TEBA 459 guaranteed re-employment to senior workers, and TEBA 458 constituted a leave certificate. By 1984, 75 per cent of workers were on one or the other certificate; in 1988, 90 per cent were.[22] Over time, TEBA 460 and 461 were phased out, and by 1990 a majority of workers were considered to be regular employees with access to leave privileges provided by TEBA 458.

By using this new system, the long-term goal of the Chamber was to have workers employed on a standard one-year contract with leave provisions. Gold Fields first introduced a nine-month contract in 1981, and extended it to one year by 1985. Rand Mines, Anglo American and JCI followed suit. By 1988 all of the mining houses used one-year contracts with a month's leave provision. Close to 70 per cent of foreign workers and over 90 per cent of domestic workers were employed on the standardised one-year contract by 1990.

In a stabilised labour market, new or retrenched workers found it much more difficult to find employment. Moreover, the recruiting of novices declined; and labour turnover and desertion were considerably reduced.[23] At the same time these processes facilitated the unionisation of the African labour force, for stabilisation generated a more durable membership base and encouraged greater commitment to workplace issues.

Preface to a Conclusion

The themes in the economic history of the gold mines discussed in this chapter form the context in which research for this book was conducted, and constitute the necessary background to the arguments that will be developed in greater detail in its two main parts. The dependence of economy and state on gold, the impact of its international price, patterns of technological change, the expansion and contraction of the labour market, and the increases in average wages were all factors relevant to the organisation of the mines' African labour force in the 1970s and 1980s – the focus of Part One, 'Organising a Labour Supply'. Similarly, the internal dynamics of the Chamber, competition and conflict between the mining houses, the mining experience and the stabilisation of the labour force were all of them processes central to the ascendance of African workers in the labour framework – which will be the focus of Part Two, 'The Ascendance of African Workers'.

What remains to be introduced here are the constraints which the political economy of gold-mining imposes on the capacities of African workers to transform its social structures. More generally, how far can change really go in modern South Africa? By themselves, the organised power of African workers in gold-mining is limited. Their union, the NUM, has made a number of considerable gains, and undoubtedly contributed to the improvement of working conditions, but they have struggled without success to remove the remaining and seemingly enduring aspects of labour repression. Migrant labour, compounds, and racial inequality in and outside the workplace remain a persisting reality, trends reinforced by the state and inter-state relations in southern Africa. The NUM has also unintentionally contributed to the retention of labour-repressive institutions by its use of the mine compound for recruiting and strike-enforcement purposes.

The NUM's answer to the structural problems of change in South African society is socialism. For the NUM, socialism has two inseparable meanings. The first is contained in the notion of workers' control or worker democracy. Union members have been encouraged to take possession of their labour processes, production, and their lives in the mine hostels. The second aspect is the nationalisation of the mining industry. The NUM has supported state-ownership of the mining industry, as originally proposed by the ANC in its Freedom Charter adopted in 1955, and reiterated by Nelson Mandela after his release from prison in 1990. Both aspects should be considered together, and not in isolation. Workers' control over production processes and the associated transformation of class structures could not happen without a change in the form and character of the state. State-ownership of the means of production would serve to reinforce what otherwise could be shortlived and tenuous systems of workers' control over production processes. For the NUM, therefore, socialism in the workplace must be cemented by a state controlled by the working class; conversely, a workers' state must be able to reinforce workers' control over the production processes.

In the conclusion to the book I consider these popular propositions about the character and prospects of social change in the mining industry and in South Africa more generally. I seek to develop three arguments. Firstly, I shall suggest that changes in the class character of the state should not be confused with changes in its racial form, and that the negotiations begun between the incumbent government and the ANC are principally about race and not class. Secondly, it will be argued that workers' control over production processes will not be possible without large-scale state assistance, specifically to alter the authority and power of management and the mine-owners. Thirdly, it will be argued that the nationalisation of the mining industry without workers' control is likely to draw state interests closer to the organised interests of capital, which could be used to curtail the power of workers during recessionary periods when work stoppages, strikes and high wage demands are not seen to be in the national interest.

PART ONE
Organising a Labour Supply

Migrant Labour and Inter-state Relations

Political Consequences of a Labour Crisis

In April 1974, the president of Malawi, Hastings Banda, ordered the withdrawal of 130 000 Malawian workers from the South African gold mines and suspended all further recruiting in his country. Himself a former worker on the mines, Banda acted in response to a plane-crash at Francistown, Botswana, in which 74 Malawian workers perished *en route* home on a WENELA-chartered flight.[1] Apparently Banda had for some time been under considerable pressure from state officials and local entrepreneurs to disengage workers from the South African mines. But whatever his reasons, his action cost the mines a third of its labour force, and the Chamber lost its single largest labour area. To compensate, the recruiting agencies turned their attention elsewhere. An agreement was concluded with officials of the Rhodesian government for 20 000 African workers, and recruiting in southern Mozambique was stepped up. The homelands were scoured for more labour, and recruiting offices opened in the country's major cities with the intention of hiring township-based workers.

In the same month, April 1974, the Spinola coup in Portugal triggered a chain of events in southern Africa that would eventually result in the independence of Mozambique and the rise to power of the socialist-oriented Front for the Liberation of Mozambique (FRELIMO) in that country. Extremely critical of the migrant labour system, FRELIMO's long-term goals were to withdraw Mozambican workers from mine employment and transform southern Mozambique – the area from which the majority of migrants originated – into a relatively self-sufficient and prosperous agricultural zone. This was seen to be the way out of a long-standing and deeply rooted dependency on South Africa. In the short term, however, FRELIMO had no intention of deliberately cutting back Mozambican labour, and, as Ruth First pointed out, were keenly aware of the dependence of state and economy on mine wages.[2]

The transfer of power in Mozambique effectively disrupted the close relationship which had existed between WENELA and the Portuguese colonial administration. FRELIMO introduced new passports, and altered the procedure by which immigration papers were to be issued to migrants. The new system took some time to implement, however. In 1976 only 4 of 21 recruiting stations were in a position to issue papers to migrants, and as a result, migration to the South African gold mines became erratic. Towards the end of 1976 few workers were passing through the system. But by the time the new immigration procedures at last became properly operational, the Chamber had in fact altered its labour policies.

From 1974 the Chamber committed the mines to a steady and firm reduction of their foreign workers.[3] Members of the Chamber now saw the large presence of foreign workers – 72 per cent of the overall labour force in 1974 – as unacceptable and dangerous: it made the mining industry vulnerable to unpredictable and changing political circumstances, it was argued.[4] For reasons like these, the recruiting agencies were instructed to limit foreign recruiting, especially in countries considered to be politically unstable. The principal targets of the new policy were Mozambique and Malawi. WENELA made it considerably more difficult for Mozambican workers to obtain employment, and kept the labour complement in check to under 30 000. In 1978 recruiting was resumed in Malawi, but employment levels were also held down, the number of workers employed in the 1980s not exceeding 20 000 per year. This contrasted with recruiting patterns in countries considered to be more stable and less likely to disrupt the labour market. Botswana had a constant 16 000 workers employed throughout the 1970s and 1980s; Swaziland tripled its labour force from 5 000 in 1974 to 16 730 by 1989; while Lesotho became the largest single foreign supply area, increasing its labour from 75 000 workers in 1975 to over 100 000, though when Basotho workers began to play a militant role in strikes during the late 1980s, Anglo American put pressure on the Chamber to limit recruiting there as well.

The power to restrict foreign labour depended on the Chamber's ability to obtain supplies from domestic areas within South Africa itself. In the next chapter I show how state policies, dependency processes and revised recruiting strategies created a buyers' market

Table 3.1. Foreign employment in mines affiliated to the Chamber of Mines, 1972–1989.

	Mozambique	Lesotho	Botswana	Malawi	Swaziland	Total
1972	80 243	67 046	19 873	106 379	4 778	278 319
1974	80 730	72 169	17 047	94 728	5 168	269 842
1976	67 439	81 973	19 871	494	9 948	179 725
1978	35 234	91 278	17 647	17 910	8 269	170 338
1980	39 539	96 309	17 763	13 569	8 090	175 270
1982	42 544	95 731	16 659	13 565	9 294	177 793
1984	44 195	95 675	17 257	15 120	10 833	183 080
1986	56 237	103 742	19 106	17 923	14 239	211 247
1988	44 084	100 951	17 061	13 090	16 171	199 783
1989	42 807	100 529	16 051	2 212	16 730	186 141

Source: Annual Report of the Chamber of Mines, 1973–1990

for migrant labour in the homelands. In consequence the recruiting agencies could select their workers from abundant labour pools, and by the late 1980s TEBA declared that it could supply all the needs of the mines with homeland labour, if need be.[5] An emergent labour market of this kind strengthened the Chamber's ability to manipulate labour supplies.

The restriction on foreign labour also required a willingness on the Chamber's part to resist the demands of labour-supplying states. For reasons rooted in their colonial past, many societies of southern Africa were firmly locked into the migrant labour system: kin networks, social systems and entire regions had become dependent on mine wages earned in South Africa. Of this more general pattern, southern Mozambique and Lesotho were extreme examples, the foundations of their very social structure resting on wage remittances.[6] For its part, Malawi had reduced its dependence in the 1970s, but nevertheless still found it necessary to send 20 000 workers to the mines. Rural communities in Botswana and Swaziland were similarly dependent on mine wages.[7]

For these foreign states the major source of revenue was derived from the system of deferred pay. FRELIMO continued a practice introduced under Portuguese colonial administration whereby workers were compelled to defer 60 per cent of their wages, which became collectable on return home. Malawi and Lesotho introduced similar schemes during the 1970s. In turn, Botswana and Swaziland had voluntary arrangements by which workers were encouraged to remit part of their wages if they so chose. The interest generated by deferred pay accrued to the various governments, not the workers; it served in effect as an interest-free credit system. In addition, supplementary revenues were generated by administrative fees charged for recruiting.

The Chamber's restriction on foreign labour therefore meant a reduction in employment opportunities and wage income for large numbers of southern Africa's population as well as losses in revenues for the affected states. For this reason, none of the state officials from neighbouring countries were willing to accept the Chamber's restrictions on foreign labour with equanimity. Even though the labour market was against them, they were prepared to put up considerable resistance.

Table 3.2. *Deferred pay and remittances to foreign countries, 1982–1988 (in South African rands).*

	1982	1984	1986	1988
Mozambique	51 102 154	70 175 699	83 462 997	102 700 000
Lesotho	111 516 891	186 281 041	240 109 132	347 800 000
Botswana	19 992 192	–	20 820 183	20 200 000
Malawi	18 857 713	26 163 962	36 878 383	54 800 000
Swaziland	9 197 429	11 446 013	12 238 577	15 200 000
Total	210 666 379	314 066 715	393 509 272	540 700 000

Source: *Race Relations Surveys,* 1982–1989.

Theories of the State

Several studies published in the 1980s highlighted the role of state officials in the elaboration of the migrant labour system and labour-repressive institutions. In his *Race and State in Capitalist Development*, Stanley Greenberg pointed to the complicity of South African state officialdom in the development of cheap labour arrangements for the mines – by instituting and administering pass laws, land-use restrictions and the homelands, suppressing labour organisation, and criminalising strikes and work stoppages.[8] In *The Emergence of Modern South Africa*, David Yudelman argued that state officials largely supported the broad labour policies of the gold mines, because by doing so they ensured an important revenue base.[9] These studies represented a considerable advance on earlier reductionist work which had tended to ignore or dismiss state institutions, by seeing them merely as expressions of dominant class interests.[10]

Considered as a whole, the literature on the state in South Africa – including Greenberg's and Yudelman's work – did not look beyond the conventional boundaries of the nation-state. Moreover, because of its hegemonic position in the regional political economy, there has been a preoccupation in the literature with the South African state and apartheid, and only passing empirical attention has been paid to the relations between states in the region. The role of neighbouring states caught up historically in the reproduction of the migrant labour system throughout the sub-continent has been ignored.[11] Partly as a result, debates about how to end the migrant labour system after apartheid have focused on the South African state and its interventions in the mining economy, to the exclusion of what it would mean for the neighbouring labour-supplying states.[12]

In her work on states and social revolutions, Theda Skocpol stressed the importance of locating state interests in revenue-collecting infrastructures, which are related to, but also autonomous from, class structures.[13] In a recent work that develops her argument, Michael Mann has traced the history of the modern state's revenue-collecting institutions, indicating their durability in a range of social orders – feudal, pre-capitalist, capitalist and socialist.[14] In support of this line of argumentation, the present chapter highlights the role of state interests and inter-state relations in the maintenance of the migrant labour system in southern Africa. By examining the responses of the Rhodesian, Mozambican, Malawian, Lesotho, Botswana and Swazi states, I shall show how post-colonial state officials continued an enthusiasm begun under colonialism in selling migrant labour-power to South Africa, and have pursued policies that maximise the revenues collectable from their workers employed in its gold mines. The role of neighbouring states – irrespective of whether they were operating in a capitalist or socialist framework – has been a critical part of the way in which the mines of South Africa organised and reorganised their labour supplies in the 1980s.

Rhodesia

Colonial Rhodesia never formed a major part of the South African recruiting network. There were many black Rhodesians working within South Africa in the 1960s, but not in the mining industry. Some 50 000–70 000 black Rhodesians lived and worked in South Africa, and in 1966 a labour agreement was signed between South Africa and Rhodesia which attempted to regulate their presence. The Rhodesians lacked documentation of their immigrant status, and the 1966 agreement sought to rectify the problem.[15]

At this time James Gemmill, head of the Chamber's recruiting agencies, became concerned about the possibility of a decline in

Malawian labour, and argued that it was worth exploring the labour potential of Rhodesia as a replacement strategy. There were problems with Gemmill's proposal, however. The Rhodesian Chamber of Mines, whose member mines had difficulty competing with South African mine wages, insisted on state protection of its labour market. As a result, the Rhodesian government made it known to the South African Chamber that it would resist any recruiting initiatives.[16]

When Malawi withdrew its labour in 1974, the possibility of hiring Rhodesian labour was raised again. The Chamber asked the South African Department of Foreign Affairs to open negotiations with the Rhodesian government. This time the Rhodesian government was receptive to the proposition. By the mid-1970s the guerilla war had escalated; unemployment in the cities, especially around Salisbury (Harare), had increased; and the government was anxious to remove the black unemployed from city streets before they joined the war. The Rhodesian Minister of Labour and members of the Department of Internal Affairs consequently gave the South African Chamber the go-ahead to begin recruiting.

The negotiations were inter-governmental. The Rhodesian Chamber was not initially consulted, probably because members of government believed (correctly) that they would have met with objections to the idea. Rhodesian mines had a tough enough time finding labour, let alone competing with South African mines. Faced, however, with its government already involved in negotiations, the Rhodesian Chamber settled for state protection of its labour market by insisting that the South Africans be restricted to urban, unemployed novices. It did not want the South African mines to pick up its experienced miners, and directed the recruiting agencies to the urban unemployed in Salisbury and Bulawayo, and then only to those with no previous mining experience.

At the end of 1974, the South African Chamber signed an agreement with the Rhodesian government for 20 000 black Rhodesian workers. This was recognised by the updating in 1976 of the 1966 inter-government agreement, whereby WENELA's right to recruit in Rhodesia was legally defined. The agreement stipulated that only WENELA could recruit 'Rhodesian Africans in Rhodesia for employment in South Africa'. Further, as of 1 April 1976, it was agreed that Rhodesians could enter South Africa to work on a 'supplementary basis in mining, agriculture and other work authorised by the South African government'.[17] The Rhodesians were signed on for either 12-month or 18-month employment terms, and had to defer 60 per cent of their wages after the first three months.

It was not a very successful experiment. Although WENELA recruited 18 653 Rhodesians in 1977, the average number hired in subsequent years proved disappointingly low. In 1978, altogether 11 984 were recruited, 7 643 in 1979 and 5 770 in 1980. The declining figures were a measure of how unpopular the Rhodesian workers became with mine management and other workers alike. They were the only urban workers in an overwhelmingly migrant-based mine culture.[18] As I shall argue later, urban workers face a number of specific obstacles posed by mining culture. On average the Rhodesians had a higher level of educational attainment, and for this reason they expected better treatment, quicker promotion and higher wages. The fact that they were usually sent to the worst mines compounded their sense of grievance. For these reasons, the Rhodesians were greatly dissatisfied, kept aloof from other migrant workers and continually subverted managerial authority.

In response, management reduced their requests for Rhodesian workers, and this was reflected in falling recruitment levels. In 1980, the new Zimbabwean government of Robert Mugabe sought to restrict its relations with South Africa to bare essentials, and expressed in public its condemnation of migrant labour as an institution. In general, Zimbabwe was not greatly dependent on migrant earnings in South Africa; unlike Mozambique and Lesotho, there was no long-standing tradition of socio-economic dependence on migrant labour. Zimbabwe revoked TEBA's recruiting licence in 1981; and by 1982, only 112 Zimbabwean workers were employed in the mines. For the TEBA organisation, it was a 'most unpopular move'.[19] For mine management it was a great relief.

Mozambique

The Chamber's reduction of foreign labour in the mid-1970s could not have come at a worse time for the emergent FRELIMO state. Within three years of independence the labour system that had once sustained Portuguese colonialism had collapsed, and Mozambique's socialist-oriented development policies failed to provide alternative employment possibilities for laid-off miners.[20] Because of the lack of jobs in Mozambique, up to 150 000 Mozambicans found employment in the 1970s on the farms of the eastern Transvaal. While the loss of employment for 80 000 workers was damaging enough, on top of it wage incomes, deferred pay and associated state revenues were reduced by some 70 per cent. Moreover, the traditional arrangement whereby the Chamber paid Mozambique its deferred pay in gold was terminated in 1978. Then, too, southern Mozambican agriculture entered a period of crisis, as a result of the collapse of colonial marketing systems. The social and economic structures of the three southern provinces – Maputo, Gaza and Inhambane – remained effectively locked into migrant labour.[21] And when the growth of RENAMO, a rebel movement supported at the time by the South African Defence Force, fuelled a civil war, agricultural recovery and economic growth were seriously undermined. Thereafter, FRELIMO began to abandon the hope of terminating its dependence on mine wages. Its workers still arrived in large numbers at WENELA's recruiting stations, yet most of them were turned away. In the light of this, FRELIMO's labour officials began to press the Chamber to change its policies, and to allow more of its workers to return to the mines.

FRELIMO's first real opportunity to bargain for increased labour came after the Nkomati Accord, an inter-state security agreement signed with South Africa in March 1983, which involved among other things economic matters and the regularisation of labour supplies. For its part FRELIMO sought to increase the annual number of workers recruited, as well as to work towards some kind of minimum quota. At the time of the signing of the accord, the supply of labour was governed by informal agreement, as the Mozambique Convention of 1928 had technically lapsed when FRELIMO came to power in 1975.

The forum in which Mozambique pressed its case was the Joint Economic Working Group, established after the Nkomati agreement was signed. The group was presided over by Mozambique's Minister of Economic Affairs and the South African Director-Generals of Manpower and Foreign Affairs, and met regularly in Cape Town and Maputo after 1983. It was in this forum that Mozambique's Minister requested an increase in the number of recruited workers to 80 000 and ultimately 120 000 a year; air transport for workers; an increased contract attestation fee; a review of the deferred-wage system; and a new labour agreement.[22]

South Africa's Manpower Minister conveyed the requests to the Chamber, and asked that they be considered prior to a full discussion between representatives of the departments of Manpower, Foreign Affairs, Co-operation and Development, and Internal Affairs. The composition of the proposed meeting was significant, as it included ministries with different views on the question of foreign labour. Traditionally only Foreign Affairs supported foreign employment, as it gave South Africa useful leverage with the surrounding states. On the other hand, the other ministries tended to be more inclined towards domestic employment policies, wanting to relieve African unemployment and to promote the government's homeland policies.

It was the Chamber, however, that most stoutly resisted the Mozambican attempt to impose quotas on the industry. Members had no intention of returning to the days when Mozambique had up to 120 000 workers in Chamber mines; they wanted to continue recruiting as many Mozambican workers as the mining houses and individual mines needed. So the Chamber rejected the request for quotas, as contrary to 'the right' of 'mine managements ... to select their own labour force in terms of skills and mine requirements'.[23] With regard to the other requests, the Chamber rejected air transportation on the grounds that it was too costly. Then again, Chamber members believed that attestation fees were an administrative cost and not a revenue-earner, implying that it would be wrong for FRELIMO to use it as an additional source of income. But they did agree to a review of wage remittances and, in principle, to a labour agreement without quotas.

While these issues were formally raised through the South African government, they were carried forward at private meetings held between the Chamber and Mozambique's Labour Ministry. From Mozambique's side, the point of the meetings was to continue to press the Chamber to lift restrictions on Mozambican labour. On its side, the Chamber wanted Mozambique to end deferred pay and to improve the efficiency of the enabling infrastructure upon which recruiting depended. In order to move the Chamber towards greater flexibility, Mozambique made a number of major concessions as far as recruiting was concerned. Its Labour Minister, A. R. Mazula, offered to lift the restrictions which limited WENELA to southern Mozambique. This restriction, imposed initially by the Mozambique Convention of 1928 – and honoured by both FRELIMO and WENELA after the Convention lapsed – was an attempt to protect central and northern Mozambique from the consequences of migrancy, and to free labour for domestic employment. But the civil war had increased unemployment and caused major population dislocation. In consequence workers from central and northern Mozambique were finding their way to the mines via Malawi. Under these circumstances, the Minister reckoned, the restriction no longer made much sense. It was therefore agreed that all reference to 26th parallel were to be eliminated from the agreement.[24]

The Minister also offered to relax the regulation that restricted WENELA to men 30 years and older. This regulation was originally introduced in support of FRELIMO's conscription system, compelling young men to complete their military service before taking up jobs. With the restriction in place, WENELA could not recruit younger novices. The Minister now claimed that Mozambique had in any case a very large labour force; unemployment was high, especially among men 16 to 26 years old; and under these circumstances, he argued that there was enough labour to serve both military conscription and mine recruitment.

The Minister also reassured the Chamber that FRELIMO state officials would attempt to discourage workers from participating in work stoppages and strikes. True to his word, at the recruiting stations in southern Mozambique, officials began to brief workers on the importance of keeping their jobs. Indeed, before the 1987 black miners' strike, Mozambican state officials made an extra effort to dissuade their workers from participating. The NUM sent a special message to FRELIMO's Labour Ministry during the strike, asking it not to send scabs to the mines.

While the Chamber welcomed these concessions, it was looking for other changes which gave Mozambique much more difficulty. The Chamber wanted Mozambique to abandon arbitrarily fixed foreign exchange rates. Secondly, it wanted to bring the compulsory deferred-pay system to an end, as it was a source of worker grievance on the mines. Thirdly, because of the war, the state could not adequately secure the safety of workers travelling through southern Mozambique. Lastly, a long-standing problem was the shortage of fresh water at Ressano Garcia, WENELA's main processing centre in southern Mozambique.[25] Tony Fleischer, TEBA's general manager until 1982, recommended in his strategic plan that the Chamber push to secure improvements in all these areas. Fleischer called for tough, even punitive bargaining and expressed the desire to exploit Mozambique's predicament to extract concessions from its government.[26]

For a country torn by an increasingly savage war, these demands were exacting. On the exchange-rate issue Mozambique prevaricated, as it did not want to abandon fixed exchange rates or set the rate too high. In the end, the exchange rate was kept at one rand to twenty metical (the rand fetched 100 metical on the black market), which was clearly to the advantage of Mozambique's fiscal offices; an exchange rate regulated simply by the market was not in its favour. Similarly, the status quo was retained on the deferred-pay issue. On the two issues the Chamber considered to be important, Mozambique succeeded in holding firm.

However, FRELIMO could do little about the deteriorating security situation in the south, having lost control over significant parts of the country. The Labour Minister claimed that (then) President Samora Machel was directly involved and promised early resolution. Despite these assurances, the matter was plainly out of state control.[27] Its request that WENELA fly workers from Maputo to Johannesburg was an admission of the security difficulties FRELIMO experienced in southern Mozambique.

Having got as far in their negotiations, the Chamber and FRELIMO were startled to learn in October 1986 that the South African government, in what both sides regarded as a perverse act, had banned the further employment and recruiting of Mozambican workers. South Africa's president at the time, P. W. Botha, had threatened on a number of occasions that he would consider implementing such a ban. He criticised Mozambique for supporting the African National Congress, and threatened to impose penalties for its support of international sanctions against South Africa. He declared that if the U.S. Congress passed sanctions legislation in 1986, he would consider passing on some of the costs to neighbouring labour-supplying states. When a land-mine explosion near the Mozambican border seriously injured a number of South African soldiers, Botha acted.[28]

Chamber members were taken aback, and became concerned about the disruptive consequences for the mines' labour supplies. Some of the gold mines were unusually dependent on Mozambican labour. At Blyvooruitzicht and Durban Deep mines, 28 per cent of the workers were Mozambicans; at ERPM, 43 per cent; at Randfontein Estates, 18 per cent; at Western Areas 28 per cent; and at Hartebeestfontein and Loraine mines, 20 per cent. In general the industry average was about 10 per cent. A variety of reasons explain the high concentration of Mozambican workers in these mines. There has been a tendency in the mining industry to send the most vulnerable workers in the labour market, who invariably were foreign workers coming from less-preferred recruiting areas, to mines with the worst safety reputation and work conditions.[29] In most of these mines, the proportion of foreign workers was unusually high. Management at the mines in question tended to favour Mozambicans because of their long-standing reputation for industriousness, loyalty to management, and indifference to unions. On these grounds, mine management made it known to the Chamber that production would be adversely affected if government was to act on its threatened embargo.

Two months after the ban was imposed, however, the government made a number of concessions. It was agreed that the mining groups could retain their long-serving and experienced Mozambicans. An agreement stated that no further novices or anyone with less than seven years' experience would be allowed to be employed or re-employed. It also stated that mines which had a disproportionate number of Mozambicans were permitted to phase them out over three employment contracts.[30] During the course of the year the Chamber persuaded government to make even further concessions, and by December 1988 the ban itself was lifted. In 1989 the government reaffirmed the Nkomati Accord, and promised greater co-operation in economic and labour areas.

These bargaining processes increased the employment of Mozambicans in the 1980s. In 1978 a total of 35 000 Mozambicans worked in the mines. By 1986 employment reached 56 000. However, Mozambique was affected by the general decline in jobs after the 1987 African miners' strike, and employment dropped to 44 084 and 42 807 in 1988 and 1989 respectively.[31]

In the 1980s, therefore, though the Chamber had not agreed to a labour quota or signed a labour agreement with the Mozambican authorities, it did relax its restrictions on Mozambican recruiting, and permitted some, if only initial, increases in employment during the decade.

Malawi

After the Chamber reduced its foreign labour component in the 1970s, Malawi was able, because of its much stronger agricultural base than Mozambique, to integrate successfully the bulk of the withdrawn migrants in the domestic economy with little disruption to its labour market. However, not all of the workers could be absorbed in this way, and Malawi still remained dependent on mine employment, albeit to a much smaller degree.[32] As a result, by 1977 recruiting was allowed to resume in Malawi. TEBA and the industry were bound to be wary, however, fearing a repetition of what they considered to be Banda's capricious behaviour in 1974.

What was critical for the mines was the need to resist any attempt on Malawi's part to impose a labour quota on the industry. In a buyers' market for labour, this was relatively straightforward. Malawi needed to send the workers much more than the mines needed to receive them. In 1982 the Governor of the Malawi Reserve Bank put a request to the Chamber for 5 000 additional workers. Although Banda claimed that Malawians no longer were interested in mine employment and that 'WENELA was dead', his office and the Department of Labour nevertheless supported the request.[33]

Fleischer at the time sought to use the opportunity to compel Malawi to remedy a range of serious problems TEBA experienced with regard to the deferred-pay system. Malawian workers deferred 60 per cent of net wages to their families through the Malawian Reserve Bank, but the bank was apparently slow in making payments. As a result, Malawian workers expressed their grievances against the Bank on the mines, and often their protest became a basis for wider industrial disturbances and union activity. Partly for this reason, TEBA wanted to introduce a voluntary system throughout the supply areas.

Fleischer suggested a trade-off. On its side, Malawi had to remedy the deferred-pay problem in exchange for a labour agreement guaranteeing jobs for 20 000 Malawians a year. In what the TEBA board of directors regarded as a diplomatic style unsuited to the industry's involvement in sub-continental political affairs, Fleischer argued that 'if a supply area does not wish to co-operate, then we must plan to phase it out of our mix'.[34] 'We must deal first of all with our friends in southern Africa', he went on to say, 'and in this way possibly influence events to our country's advantage.' Delivering a mild rebuke to Fleischer, the TEBA board noted that while the organisation 'might wish to take up with the governments of the countries supplying labour to the industry specific issues ... from time to time, it should nevertheless not change its public stance towards such governments but should continue to conduct itself towards them using low-key diplomacy.' The claim affirmed the continued preference for privately conducted, rather than publicly declared, negotiations. The TEBA board of directors declined to guarantee employment to 5 000 more workers as Malawi requested, on the general principle that 'no quotas were set for any labour supplying country'.

The growing incidence of AIDS in Central Africa was to make a major and largely irreversible impact on recruiting practices in Malawi.[35] At first, the Malawi government denied, then understated, the high levels of AIDS in the region. Only by 1986, after the World Health Organisation had established that Malawi had a 6 per cent HIV infection rate, did government issue an official pronouncement, and then only referred to it anonymously as 'this urgent, serious, health problem'.[36] No case figures were provided. In September 1986 government turned to local herbalists to assist with finding a cure for AIDS, while at the same time launching a public awareness campaign, whereby the public was advised that a 'change in our attitude towards sexual relationships' would be required to combat the disease.[37] By February 1987, no longer able to sweep the issue under the table, an anti-AIDS committee was formed to 'educate the public on the problem of AIDS'.[38]

Fearing the spread of infection in the mines, the Chamber began testing for AIDS among miners from Malawi and other high-risk areas. (There were occasional AIDS cases reported in Lesotho and Botswana.)[39] The Chamber wanted to test for AIDS in Malawi, but state officials turned down a request to do so. The South African government, in turn, put enormous pressure on TEBA to repatriate all identified AIDS carriers to their countries of origin. Legislation was passed in 1987 making this possible.[40] The Chamber thereafter only re-employed experienced Malawian miners and refused to recruit any novices. As a result, the employment of Malawians declined over a period of three years from 17 939 workers in 1987 to 2 212 by 1990.[41]

Lesotho

When the labour crisis on the gold mines first broke in 1974, Lesotho was well placed to take up the slack created by the withdrawal of Malawian and the shortage of Mozambican labour. Lesotho had for decades supplied the mines with labour. With the development of the Orange Free State mines in the 1950s, labour supplies expanded, and by the early 1970s some 70 000 Basotho workers were employed annually in the mines, making up 20 per cent of the mine labour force. Almost half the workers were employed in the Free State, in Anglo American-owned mines.

As in southern Mozambique, the mines had become a major source of employment, income and revenue for society and state. In the 1970s, about 40 per cent of gross national product came from mine earnings; in the 1980s it grew to 50 per cent.[42] Unemployment was high in Lesotho, and as a result the demand for mine employment great: 'hundreds of eager, potential workseekers ... are turned away each day', the TEBA manager based in Lesotho's capital, Maseru, reported.[43]

Leabua Jonathan, Lesotho's prime minister till 1986, encouraged increased mine employment as a source of revenue. Employment grew from about 70 000 Basothos hired in the mid-1970s, and reached over 100 000 by the mid-1980s. With the eclipse of Mozambique and Malawi as suppliers, Lesotho became the largest foreign source. Jonathan sought to maximise state revenues from mine employment. Indeed, in order to increase the state's stake, he rushed the Deferred

Pay Act through parliament in January 1976. This required the repatriation of 60 per cent of total mine earnings through the Lesotho National Bank. Aware of the instability in the other supply areas, Jonathan felt that he could risk taking this step.

But the Deferred Pay Act was unpopular with Basotho workers. They disliked the fact that they had to forcibly surrender a part of their wages to a government they did not trust. J. K. McNamara has underlined the role of worker grievances against the deferred-pay system in his study of inter-group violence on the mines in the 1970s.[44] What is more, Basotho workers became eager to join the National Union of Mineworkers because they believed that the union represented a vehicle by which pay-related grievances could be redressed. In fact, Basotho workers became major participants in work stoppages and strikes in the 1980s. In response, members of the Chamber complained that Basotho workers expressed their grievances against deferred pay on the mines. Mine management, and not the Lesotho government, had to live with the consequences of the deferred-pay system, they argued.

Personally, Jonathan resented the migrant labour system, and saw it as demeaning and dehumanising; and in general he sought to keep his distance from South Africa and apartheid. The South African state, on its side, supported opposition groups within Lesotho, and evidently tried to subvert Jonathan's rule. In two known cases, the South African Defence Force raided Lesotho under the pretence of hunting down African National Congress guerillas, and there were ongoing struggles between the two countries over the policing of borders. For their part WENELA officials complained that they had enormous administrative difficulties with the government. Not only did the state radio broadcast anti-WENELA messages regularly, but Fleischer was so incensed about the treatment WENELA officials received in Lesotho that he wanted to turn it into a bargaining issue.[45] Fleischer's successor, Errol Holmes, spoke about Jonathan's attitudes as one of the major problems the agency had with Lesotho, noting that the administration used Basotho mine labour earnings 'unashamedly ... for their own political ends'.[46] WENELA's objections to the Jonathan administration were summed up by the district manager based in Maseru: 'TEBA's position in Lesotho in the past was tolerated by the politicians, who while realising the importance of our operations to this country never really accepted us willingly and used every opportunity to squeeze us for every cent they could.'[47]

In January 1986 Jonathan was toppled in a coup d'état, and was replaced by Major-General Justin Lekhanya as the new head of state.

Lekhanya maintained Lesotho's commitment to migrant labour and the compulsory deferred-pay system. Though the Chamber pressed to have it changed to a voluntary one, arguing that significant revenues could be raised in this manner too, Lekhanya saw no merit in altering a system that served the state so well. After a number of failed attempts to change his mind, the Chamber's Technical Advisory Committee decided 'to let sleeping dogs lie'.[48]

Unlike Jonathan, Lekhanya proved much more willing to establish an openly sympathetic relation with the Chamber and its recruiting operations. TEBA's district manager in Maseru reported to head office in 1986 that the situation 'has changed completely'. TEBA operations in the country were 'now fully accepted and indeed welcomed'. He was pleased with the 'direct access to the Labour Minister himself', and that relations between WENELA and the Labour Ministry had improved considerably. In order to cement the emergent cordial relationship, the district manager suggested that the Chamber donate the existing recruiting facility in Maseru, the Kingsway property, to the Labour Ministry. He noted that the government was broke, that TEBA needed a much larger facility, and that the gesture would not be lost on the government.

By the late 1980s, African trade unionism and worker militancy began to make an impact on the recruiting strategies of the Chamber. As noted earlier, Basotho workers developed a keen interest in the NUM, hoping that the union could remedy the compulsory pay-deferral system. They also at first believed that membership of the union would serve as a guarantee against mass dismissals and retrenchment. Jonathan himself had supported the processes of unionisation on the mines, seeing it as a means of eroding apartheid and empowering black people. As McNamara has recorded, during the early phase of unionisation the NUM concentrated initial recruiting on senior mine workers, assuming that this would bring along the junior workers. Because Basotho workers dominated senior positions in Free State mines,[49] they came to play a prominent role in the development of the NUM.

By 1987 union membership and participation in work stoppages had become a liability. Lesotho state officials feared that the Chamber would come to view Basotho workers negatively. Lekhanya – who took power at a time when the NUM was gearing for two years of militancy – became concerned about the role of Basotho workers in strikes and work stoppages. From the state's point of view, it was vital, even patriotic, for all Basothos to keep their jobs: 'The new government has stated categorically', TEBA's district manager reported after the coup, 'that the Basotho on the mines should do everything in their power to protect their jobs.'[50] Just prior to the miners' strike of August 1987, two Lesotho ministers met with a mining delegation, and stressed 'the importance to Lesotho of maintaining its nationals in employment on the mines'.[51] During and after the strike, senior Lesotho officials visited the mines and tried to dissuade the workers from participating.

As it turned out, the workers paid little attention to the state, and the participation of Basothos in the 1987 miners' strike was significant. Fearing that the Chamber and mine management might reduce their overall employment, delegations from the Lesotho government – including the Ministers of Labour and Interior – and the mining industry met, and the mines were asked to maintain Basotho employment at pre-strike levels.[52] A further delegation met with the president of the Chamber and Peter Gush of Anglo American to discuss the effects of the strike on Lesotho and its people.[53]

In the immediate aftermath of the strike, the WENELA district manager reported 'strong indications of a shift away from Basotho labour' in requisitions.[54] In the labour orders telexed to TEBA during the strike, Anglo American mines specifically asked for workers from Mozambique, KwaZulu and Bophuthatswana, and not Lesotho. After the strike, Anglo American, the largest employer of Basothos, decided to limit their employment. Basotho and Xhosa workers from the Eastern Cape were seen as largely responsible for the strike, especially for its violent and militant character. Though the reduction of Basotho labour would be left to individual mines, there was a general feeling that Anglo American should reduce its dependence on Basothos in the 1990s.

Botswana and Swaziland

Throughout the 1970s and 1980s, the Botswana government was anxious that the Chamber might reduce employment levels. The president of Botswana was therefore dismayed when TEBA closed down its northern operations in 1982, as part of its regional rationalisation of services. In order to guarantee employment levels, Botswana's president requested in the early 1980s that the Chamber hire a quota of 20 000 workers annually. The Chamber refused to sign a written agreement to that effect, but agreed verbally to do so. The agreement was considered to be 'good diplomacy', in so far as it was recognised that 'the arrangement was of greater value to them, Botswana, than to the industry'.[55]

In 1986 TEBA's district manager in Swaziland made the now familiar comments about unemployment levels, a high demand for work among Swazis, and the ease with which recruiting was proceeding.[56] Officials in the state were concerned about maintaining employment levels in the mines, and hoped that they could be increased. Between 1974 and 1989, Swaziland tripled its labour supply, from 5 168 to 16 730 workers. State officials were pleased about the increased labour requisitions made during the 1987 strike. Prior to the strike, the Swazi Minister of Foreign Affairs personally offered the services of his government if the TEBA district manager needed it. At the occasion of the opening of a new factory, King Mswati 'condemned all strikes' and assured the audience that his government would do a great deal to 'maintain cordial industrial relations' in the mines.[57]

The Southern African Labour Commission

In an attempt to promote co-operation at a policy level between the labour-supplying states, the International Labour Organisation (ILO) sponsored a number of meetings at which the idea of a migrant labour cartel was discussed. In 1977 state officials met in Lesotho to formulate the rudiments of a co-operative policy. In 1978 a conference sponsored by the United Nations and ILO deliberated on the feasibility of a common labour policy.[58] In 1980 Botswana, Lesotho and Swaziland agreed to form a Southern African Labour Commission. Mozambique and Zimbabwe later became full members, and Malawi and Zambia gained observer status. The Labour Commission was to meet once a year to discuss jointly and, where feasible, act on common concerns in the migrant labour market.

Although there was initial enthusiasm for the complete phasing out of migrant labour, especially among those countries least dependent on its revenues, it soon became clear that this was not in the interests of the states concerned. Every single state official raised moral and political objections to migrant labour, but all the same remained acutely conscious of the reality of enduring dependence. Advisors from the ILO noted that if migrant labour was to be terminated, it should be phased over a period of time, such as fifteen years or so. It was further suggested that the phasing out should be accompanied by employment-creating development projects in the sending countries, so that returning migrant workers could be productively absorbed in domestic employment.[59] In the short term, the ILO advisors encouraged state officials to extract as many benefits as possible from the migrant labour system.

Though by the 1980s the Labour Commission would perform an important watch-dog role over conditions of employment and workers' rights, it was unable to put into practice any of the proposals regarding the withdrawal of foreign workers. As a labour cartel, the Commission frankly failed. Labour migration increased in the 1980s when reductions should have occurred. Furthermore, the existence of the Commission could not stop individual states from acting on their own, negating the co-operative initiative.

Conclusion

This chapter has examined the role of South Africa's neighbouring states in the reproduction of the migrant labour system. During the colonial period, these states had developed revenue-collecting infrastructures based on the sale of migrant labour to South Africa's mines. In the post-colonial period, the dependency of state and society on mine employment in South Africa persisted, and state officials pressed for even greater involvement of their workers in the mines' migrant labour system. This was as much true for those states working within a capitalist framework, as for those pursuing a socialist one. Dependency went well beyond the particular economic frameworks within which the states of southern Africa operated.

The dependence of labour-supplying states on mine employment, and their enduring interest in maintaining the migrant labour system, is one of reasons why migrancy may continue into the post-apartheid period. The pressure these states will exert on South Africa to maintain the status quo is unlikely to diminish. For their part, the African National Congress and other black political parties will find it difficult to resist the requests of those states that supported them during the decades of struggle against apartheid, even though the support was often inconsistent and symbolic. It will be an extremely difficult and certainly a long-term task to undo the structural relations of dependency which the migrant labour system has generated over the hundred years of its existence. The end of apartheid and the changing character of the state will not necessarily change those relations.

4

A Buyers' Market: Labour Recruiting in the Homelands

Back to the Homelands

In the 1960s and early 1970s, the Chamber hired a quarter of its African labour force for the gold mines from within South Africa. Most of the workers came from the Eastern Cape, specifically the African homelands of the Transkei and Ciskei. In the face of the labour shortages of 1974–5, members of the Chamber decided to expand labour recruiting to other domestic areas. Recruiting offices were opened in Johannesburg, Cape Town, Durban and Port Elizabeth in 1974, and an agreement was signed with state officials to enable the Chamber to experiment with the recruiting of farm labour in four agricultural districts. The urban recruiting initiatives failed to deliver workers on a significant scale, however, and as a result the offices were closed down by 1976. As for the recruiting of farm labour, this was a restricted exercise, and indeed the Chamber did not expect large numbers of workers to come from this source.

The major focus of expanded domestic recruiting became, therefore, the homeland areas. This was neither new nor surprising. In the 1920s and 1930s, when high mortality rates among workers from tropical Africa resulted in a decline of foreign recruiting, the mines turned to the reserves in South Africa for their needs.[1] But during the late 1930s and 1940s, the industry began to lose local workers to better jobs in the growing manufacturing and service sectors stimulated by the war effort.[2] In the face of this, the Chamber reverted to foreign countries for the bulk of their labour, and by the 1960s workers from Malawi and Mozambique dominated the mines.[3] When the Chamber once again turned to homeland labour in the aftermath of the foreign labour crisis of 1974–5, it continued, therefore, a well-established

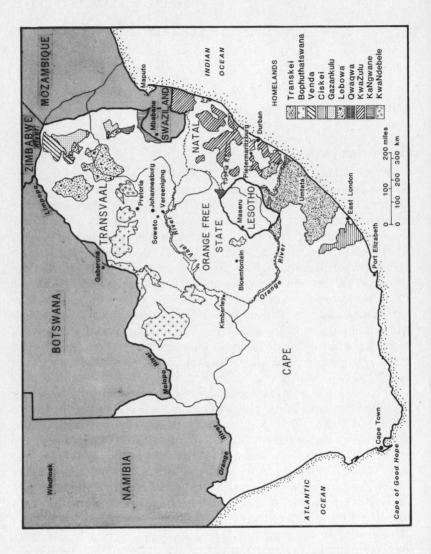

historical pattern.

New was the Chamber's interest in a number of non-traditional recruiting areas within South Africa. Chamber members were very keen to expand recruiting operations in the homelands of the Transvaal – especially Bophuthatswana – and Natal's KwaZulu. Those mines administered by Anglo American Corporation developed a special interest in attracting labour from the homelands and rural townships of the Orange Free State. As the success of these recruiting efforts became evident, the Chamber began to limit the flow of African labour from its traditional sources in the Eastern Cape. For one thing, Chamber members were anxious about their over-dependence on a single source area and consequent vulnerability, especially in the light of their recent experience with Malawi and Mozambique.[4] Moreover, they became concerned when workers from the Transkei began to feature prominently in NUM-led strikes in the 1980s. Anglo American, the largest single employer of Transkeian labour, pressed the Chamber after the 1987 miners' strike to limit the employment of workers it referred to as 'Mandela's children', from the Eastern Cape.

The ability to expand labour markets in the homelands was related to dependency processes, recruiting incentives, and the state's apart-heid policies. A number of recent studies have indicated how deteriorating conditions in the homelands affected migration pat-terns to the mines.[5] Though variable, the degree of access by homeland populations to a means of subsistence had eroded consid-erably by the 1980s. In particular the relocation and resettlement policies of the South African state, whereby Africans were expelled from white farms, 'black spots' and cities under the influx control system, accelerated the process of dependency. The size and density of homeland populations grew and unemployment increased; vil-lages in the homelands now contained people with declining access to an independent means of subsistence; and a proletariat emerged in the proliferating rural slums as a consequence of state policy.

Dependency processes were not on their own sufficient to generate the desired level of labour supplies for the mines. While unemploy-ment and poverty were powerful incentives to find work, they did not necessarily ensure that workers would turn to mine work. As the Chamber realised, work-seekers had to be drawn actively into the labour market. The Chamber thus began to improve the incentives available to recruiters. Shorter employment contracts – of 26 and 48 weeks – were offered in place of the longer 52-week ones. (However, when it became evident that a buyers' market for labour was emerging, the employment contracts were later lengthened to 52 weeks.) Improved wages were also offered to attract homeland workers. Whereas in the 1960s and early 1970s, African wage rates had been kept very low, thereafter the improved selling price of gold made it possible for the Chamber to make adjustments to wages, which it did with the major increases introduced between 1974 and 1980. The principal motivation of the improved wages was to attract more homeland-based workers to the mines.[6]

Table 4.1. Domestic workers employed by mines affiliated to the Chamber of Mines, 1983–1989.

Recruiting offices	1983	1985	1987	1989
Transkei	126 511	138 994	135 655	114 893
Bophuthatswana	27 808	39 128	46 676	–
Orange Free State*	24 523	26 792	32 696	29 995
KwaZulu	21 317	21 834	38 755	39 358
Transvaal*	8 843	10 611	28 091	29 472
Ciskei	14 434	13 788	14 082	12 000
Cape*	14 646	13 616	11 879	9 085
Qwaqwa	5 922	7 093	8 876	7 912
Lebowa	7 231	7 472	13 479	10 756
Gazankulu	2 803	3 226	6 331	5 594
Natal*	3 311	3 192	9 126	7 779
Venda	2 214	2 584	2 296	2 002
Total	263 588	288 330	316 825	281 080

* Depots
Source: 'Summary of monthly field reports', TEBA, 1984–1989.

The workers responded by pouring into the labour market. By the late 1980s, the managing director of TEBA calculated that all of the mines' labour needs could now be met from the homelands alone.[7] If forced to do so, the mines could in fact survive without foreign workers, TEBA officials believed.

The flooded labour market had a number of consequences. The recruiting agencies could now pick and choose their labour. The most desired workers were placed on special contracts, and encouraged to stay in mine employment over longer periods of time. Then again, new and retrenched workers found it increasingly difficult to obtain employment. As a result, the functions of the recruiting agencies began to change. In a sellers' market, with workers in short supply, the TEBA offices had to search actively for labour. In a buyers' market, the offices processed career-oriented workers in possession of one or more of the industry's guarantees of re-employment. The high level of worker stability that flowed from this effectively transformed the administrative character of the mines' labour bureaucracies.

Reproducing a Labour Force

In his theoretically driven essay on migrant labour systems in southern Africa and California, Michael Burawoy made the useful distinction between the 'maintenance' of a labour force and its 'renewal'.[8] 'Maintenance' processes sustain the individual worker, while 'renewal' processes create the next generation of workers. Burawoy argued that both processes are part of the way a labour force is 'reproduced' over time. His study noted that the 'cheapness' of mine labour in South Africa derived from the fact that 'renewal' functions were performed in the homelands, away from the formal Republic of South Africa. By keeping the families of migrants in the reserves, the employers and the state were absolved from the costs of supporting what otherwise would be a much larger population. Thus, mine wages 'maintained' the African worker at low cost, while the household economies 'renewed' the worker's family in the homelands.

Our understanding of the specific nature of 'renewal' in the homelands involves, of course, largely empirical questions. Harold Wolpe (and Chamber members whose statements and information he widely used as evidence)[9] argued that subsistence production 'renewed' the migrant worker, by providing a supplement to the wages earned for 'maintenance'.[10] Wolpe believed that given the erosion of subsistence production in the reserves in the 1930s and 1940s, state officials tried to reorganise the politics and economics of the homelands so as to reinvigorate their 'renewal' functions, under a programme of apartheid. Charles Simkins and other scholars have questioned the empirical veracity of Wolpe's arguments, pointing out among other things that subsistence production had already declined in the 1910s and 1920s, well before apartheid emerged.[11]

In their study, *Uprooting Poverty: The South African Challenge*, Francis Wilson and Mamphela Ramphele recorded that by the 1980s mine wages had become the most important source of income for families of migrants in the homelands. Based on data from the Transkei, they noted that of those family incomes of less than R1 500 (US$600) per year, 66–71 per cent came from migrant wage remittances; 14–19 per cent came from pensions; and 11–15 from wages in local jobs. Only 2–3 per cent came from home and subsistence production.[12] Although there were some differences between one homeland and another, Wilson and Ramphele's findings for the Transkei are broadly applicable to other homelands and have been corroborated by other researchers.[13] Corporate research confirmed that by the 1980s only a fraction of migrants were concerned about or had interests based on access to land and livestock in the homelands.[14]

These figures indicate the relative insignificance of subsistence and household production in the reproduction of the mine labour force in the 1980s. Instead of subsidising mine wages, families of migrants depend in fact almost entirely on outside sources of income, particularly the mine wage, for their livelihood. Moreover, because of the high levels of unemployment in the homelands and the stabilisation policies of the mines, jobs in the mines became scarce and sought after. Whereas in the past mine jobs had been taken up as last resort, now they became preferred avenues of employment. TEBA officials reckoned that mine jobs had become as good as 'gold' in the homelands.

A number of authors have suggested that 'renewal' was not so much secured by subsistence production, as by the role of women in the household economy.[15] In an unpublished work, Amelia Mariotti was first to argue that 'renewal' functions in the migrant household rested with the material activities of women, whether or not they were involved in subsistence agriculture.[16] In his recent review of the historical and social science literature, Robin Cohen took the argument further by referring to adaptations and adjustments in the gender division of labour, and the changing role of women in peasant households. Cohen argued that the peasant household 'subsidised' the 'capitalist mode of production', even when subsistence production was no longer viable. Women were said to have a 'special burden' because of their involvement in the 'biological reproduction' of labour-power. They had to cook food and look after the family's health, clothing and housing needs; children had to be nursed, educated, trained and socialised. Peasant households became 'points of refuge, rest and recuperation for exhausted labourers', and when sickness or injury struck they became 'field hospitals'. During periods when workers were laid off or retired, the peasant household provided the equivalent of 'unemployment benefits, redundancy payments and a pension'.[17]

Cohen's analysis reads more like a research agenda rather than a series of propositions grounded in empirical work. Little by way of original research material has been offered by him against which his more theoretically derived propositions can be evaluated. But whether they were concerned with subsistence production or with the role of women and gender relations, the 'renewal' literature focused attention and questions on workers already in employment.

The overriding concern in this literature has been to give an account of the social, economic and political mechanisms that 'reproduce' individual workers with jobs, with no attention paid as to how workers gained access to jobs in the first place, or how the structure of the labour market made access to jobs difficult.

For the overwhelming feature of workers' experience in the homelands in the 1980s has been the systemic lack of access to jobs. This has been the result of changes in the political economy of the homelands brought about largely by apartheid policies, and specifically by processes of mass resettlement and removals. The chronic lack of access to jobs created a buyers' market for labour, so that there were many more workers available than there were jobs. The mines could therefore be very selective about their labour, hired from a wide variety of often competing labour source areas. What the mines did was to reproduce their labour force by manipulating supplies, by turning the labour tap on in some regions and off in others.

The Role of State Officials

In their study for the Surplus People Project, Laurine Platzky and Cherryl Walker documented the role of the state in the resettlement of African populations.[18] They noted that in pursuance of the state's apartheid policies, close to 3.5 million Africans were moved to the homelands between 1960 and 1983. At the time of their writing, another 2 million people were still under threat of removal. The largest category consisted of people from white farms, where as a result of the combination of mechanisation and rationalised labour policies workers faced bleak employment prospects. And for its part, fearing the *verswarting* (blackening) of the countryside, the development boards of the Ministry of Co-operation and Development removed displaced farm workers to the homelands.

Resettlement changed the regional distribution of the African population. Charles Simkins estimated that in 1960 about 30 per cent of the African population lived in the cities, 31 per cent in rural areas, and 39 per cent in the reserves. By 1980, 27 per cent lived in the cities, 21 per cent in rural areas, and 53 per cent in homelands.[19] If one allows for natural increase, there was a 14 per cent growth in the proportion of the African population residing in the homelands between 1960 and 1980, due mostly to state resettlement.

According to Platzky and Walker, four types of homeland settlements emerged: border towns for commuters; rural townships for families of migrants; closer settlements for squatters evicted from farms and mission stations; and those for squatters from agricultural settlements. In 1980 some 55 per cent of the population in all the settlements and 37 per cent from rural townships and closer settlements were migrant workers. In the rural townships, 28 per cent of the population were unemployed, 22 per cent in closer settlements. These figures were considerably higher than the already alarming unemployment rate of South Africa's urban African population.[20]

Confronted by these demographic patterns, state officials appealed to the Chamber to hire more workers from a context the state had largely created. Already in the early 1960s, when foreign employment in the mines was at its height, the Department of Bantu Affairs asked the Chamber to consider hiring more workers from the reserves. In the 1980s, its successor, the Department of Co-operation and Development, renewed the long-standing effort of the state to promote the homelands as preferred areas of mine employment. Homeland governments themselves appealed directly to the Chamber to hire more of their workers for the mines. They also put pressure on the Department of Co-operation and Development officials to formulate a policy that favoured homeland over foreign workers.[21]

While the Chamber's turn to homeland labour in the late 1970s was welcomed by state officials, they insisted that it did not go far enough. An official letter to the Chamber sent in 1981 disapproved of the 'high number of foreign blacks in the mining industry' and threatened that the issue was receiving urgent attention at the 'highest levels' of government.[22] The director-general of the Department of Co-operation and Development noted that a number of junior officials in the field were complaining that 'local workseekers cannot find work on the mines due to the fact that foreigners are recruited for work'.[23] His department felt that the hiring of homeland workers should be a 'high priority', since 'thousands of unemployed blacks' could be trained for work in the mining industry. He urged, therefore, that the dependence of the mining industry on foreign black migrant labour 'ought to be reduced'. In addition, the Chamber was asked to account for the 'steps being taken' to hire more homeland labour.[24]

The Chamber's response to state pressure was to insist on a strong version of managerial prerogative in the allocation of labour. Members claimed for themselves the right to 'select, train and reward' individuals in terms of the mines' 'own employment policies'.[25] A senior official of the recruiting organisations noted that the mines needed 'a balance of labour skills', and therefore it was important to have a 'free and wide choice in selection'. As a fundamental principle, the official declared, 'outside parties' should not interfere in 'employment requirements'. The Chamber, he went on, had already increased the number of homeland workers in the late 1970s, and saw no reason why the established 'labour mix' should be changed to suit state-defined ideological ends. What was crucial for the Chamber, however, was the need to reorganise the domestic labour market – to expand recruiting in some areas and to restrict it in others.

The Mining Houses and the Homelands

As the body representing the interests of all mining employers, the Chamber developed in the mid-1970s the practice of hiring more homeland workers as the general policy of the industry. However, among individual mines and mining houses there was an uneven commitment to this policy, and an uneven capacity to follow the Chamber's guidelines. All of the mines had some foreign workers in their labour forces, for reasons to do with maintaining a large number of competing source areas, and because of the market discipline this imposed on domestic workers. More specific explanations have to be advanced, though, for the differences between the mining houses regarding the relative balance between foreign and domestic labour. Statistically, in the 1980s Gencor drew 28 per cent, Anglo American and JCI each 40, Gold Fields and Anglovaal 50, and Rand Mines 60 per cent of their labour forces – some 200 000 workers in all – from foreign source areas.

One reason was the mines' reliance on worker groups from foreign source areas for sought-after skills. Because they had worked in the mines and dominated skilled work for so long, foreign workers had come to specialise in certain tasks. Basotho workers, for example, specialised in shaft-sinking operations, and carved out an occupational niche in this line of work.[26] Mozambican workers, on the other hand, developed a reputation for being exceptional machinists.[27] Over time, cultural and regional stereotypes like these emerged, according to which certain worker groups by virtue of their nationality were associated with particular jobs. The mines' hiring practices and the custom of workers' encouraging recruitment of their own kind combined to reproduce what became internal ethnic labour markets.

Another reason for the reliance on foreign labour was the relative distance between the place of work and home for individual workers. It was often advantageous for individual mines to hire workers from proximate areas. Time and expenses involved in travelling from home to work and back could in this way be reduced. Workers could return home and spend time with their families over weekends when they were free, instead of remaining in the hostels. Management believed that this promoted worker stability, labour peace, and improved productivity. The mines in the Orange Free State, for example, hired most of their workers from Lesotho, Qwaqwa and local townships, because of the relative ease of labour mobilisation.

The most compelling determinant, however, of the relative balance of foreign as opposed to domestic workers in any given mine was its safety reputation. Through the informal network of communication that stretched from mine to mine, workers could tap a common body of knowledge about prevailing conditions of work throughout the industry. They generally tried to avoid working at the worst mines. Management at those mines with low safety ratings had great difficulty attracting and retaining their labour forces. Two sources of statistical information reflecting the safety reputation of mines can be cited to illustrate the point: the reportable accident statistics of the Chamber, and the level of novice recruitment. The correlation between these statistics and 'labour mix' indicates that at those mines with a poor reputation, the concentration of foreign workers was especially high.[28]

Typically, management at poor mines turned to foreign workers to solve their labour problems. Foreign workers were more vulnerable in the labour market: state immigration policies made it much easier to manipulate labour supplies and to retrench foreign workers. Moreover, the demand for jobs was very high, and workers could easily be replaced. As a result, foreign workers were sent to the worst mines, where domestic workers by virtue of their greater security in the labour market preferred not to work.

In sum, dependence on particular skills, the relative proximity of a mine to its source area, and the mine's safety reputation, all entered into the foreign–domestic equation. Typically, the worse the conditions on a mine, the more difficult it was to reduce the presence of foreign workers and replace them with domestic workers. In the 1980s an additional factor influencing the labour ratio was how well represented the NUM became in particular source areas. During the 1987 miners' strike, for instance, management began to hire fewer workers from areas where the NUM was strongly represented.

Expanding Supplies I: Transvaal

In the 1980s, the homelands of the Transvaal – especially Bophuthatswana, Lebowa, Gazankulu and Venda – became the second most important area of recruitment for the Chamber. Of these, the most important was Bophuthatswana, an area from which the mines had not previously recruited labour in significant numbers. It was the development of Bophuthatswana's platinum mines, as well as more intense recruiting for the Transvaal's coal and gold mines, that was responsible for increased demand.

Like other homelands, Bophuthatswana was deeply dependent on the South African state for the bulk of its revenues.[29] It had a mining, quarrying and a small number of manufacturing industries, but these contributed only a quarter of its gross domestic product. Moreover, the ability of the regional economy to generate jobs for Bophuthatswanan residents was poorly developed. During the recession of the 1980s, commuter workers employed at car assembly plants located close to Pretoria were retrenched: BMW, Sigma Motors and Siemens industries laid off over a thousand Bophuthatswanan workers. The asbestos mines had also retrenched workers. And four successive years of drought added to poor economic performance.

High levels of unemployment became a persistent problem for the homeland. TEBA's district manager reported to head office that the unemployment situation throughout Bophuthatswana was 'serious' and had 'escalated' to 'alarming proportions'.[30] He saw no signs of improvement in the immediate future, noting that 'any type of employment' was in great demand. More workers were offering their labour than could be hired, and many had to be turned away.[31]

These perceptions of the labour market reinforced optimistic conceptions among TEBA officials in the field; the labour prospects here were regarded as 'very good'. In the 1970s, the gold mines of South Africa employed about 15 000 workers from Bophuthatswana; by the late 1980s this had expanded to 47 000. Other homelands in the Transvaal increased their labour supplies as well. Labour from Lebowa increased from 7 231 in 1984 to 10 756 by 1989; and from 2 803 to 5 594 in Gazankulu. Only Venda's remained at the small figure of 2 000. Workers hired at TEBA's depots in the Transvaal, but originating from the various homelands, jumped from 8 843 in 1983 and reached 29 472 by 1989. All told, workers from the Transvaal area doubled from 46 889 in 1984 to 93 821 by 1989, and it became the second most important region in the domestic recruiting network. In 1989 a fifth of the mines' labour force came from the Transvaal alone, and a third of all domestic labour.

Expanding Supplies II: The Orange Free State

In the 1970s, the mines began to show a growing interest in hiring African workers from the homeland of Qwaqwa and the rural African townships of the Orange Free State. Both homeland and townships were products of the state's resettlement policies. Qwaqwa's population had increased from about 25 000 to under half a million between 1970 and 1984 as a result of resettlement. Of those moved a majority were workers from white farms in the province, displaced because of mechanisation, changed conditions of employment, the loss of land and grazing rights, and the erosion of labour tenancy.[32] Many of these displaced farm labourers also found their way to Botshabelo, a rural slum adjoining one of the fragments of land that make up Bophuthatswana, fifty kilometres east of Bloemfontein. This township came into being when the state removed some 38 000 Sotho-speaking people from Thaba 'Nchu in 1979 to a former farm called with unintended irony Onverwacht (Unexpected), later renamed Botshabelo. Labour tenants evicted from white farms and 'deproclaimed townships' were subsequently moved there too. By the late 1980s Botshabelo had a population of over 200 000 people, becoming the second largest African township in South Africa after Soweto.[33]

The proximity to the Free State mines made labour from Qwaqwa an attractive proposition. Some of the workers could commute to the mines, and visit their families more often. When there were special orders, the TEBA offices were able to provide the labour at short notice. Moreover, worker dissatisfaction over deferred pay (remitted only to foreign states and some homelands) was avoided. Anglo American, the largest employer of labour in the Orange Free State, wanted to hire more commuter and urban-based workers, and sought to create housing estates for more senior African employees living in the existing townships.

Recruiting in the Free State was underscored by high levels of unemployment. The Free State TEBA manager wrote that unemployment was 'the biggest problem in the area'.[34] In Botshabelo it had reached 'disturbing' levels. Other jobs were not easily available in the area and there were many more workers offering their services than could be hired by the mines.

Recruiting prospects in the 1980s were, therefore, considered to be 'very good'. The Qwaqwa area was fast developing into a 'stable' supply area. By the late 1980s, some 8 000 African workers were hired from Qwaqwa, and another 30 000 from African townships and other parts of the Orange Free State. Thus in the course of one decade, the Orange Free State had become a significant area of labour recruitment for the mines.

Expanding Supplies III: Natal

In the 1980s, the homeland of KwaZulu, scattered in unconsolidated fragments of land throughout the province of Natal, became the third most important area of expansion for the recruiting agencies. Previously, the mines had hired a small number of workers from KwaZulu. In the 1980s, however, social conditions in the homeland favoured a sizeable expansion of recruiting. The regional economy was not generating enough jobs, and unemployment levels were high. The KwaZulu district manager wrote to TEBA's head office that unemployment in his area of operation had become 'frightening', and that it could only worsen unless that was a pick-up in the economy.[35] He noted that few employment opportunities were being generated in agriculture, industry, or building and construction. Because of the large numbers of workers who could not find work anywhere, the mines' prospects for labour seemed 'better than ever'.[36]

The offices of the KwaZulu homeland government supported TEBA's recruiting initiatives, and were willing to assist where possible. Gatsha Buthelezi, chief minister of KwaZulu, appreciated the 'generous intake of Zulus' in the mines, and pledged his government's support for further expansion of recruitment. Although critical of the migrant labour system, Buthelezi saw mine employment as an answer to persisting unemployment problems in Natal and KwaZulu. He was grateful to the mines 'for what they do for hundreds of thousands of black people'.[37]

In 1985, the mines recruited 21 834 workers from KwaZulu; this figure almost doubled to 39 358 by 1989. Given the flooded labour market, a majority of these workers wanted long-term employment guarantees, and thus signed up under one or other of the industry's re-employment certificates. What is more, the high percentage of novices recruited indicated TEBA's intention to enlarge labour pools from KwaZulu.

Restricting Labour: The Eastern Cape

In the immediate aftermath of the foreign labour crisis of the 1970s, the Eastern Cape became the single most important area of recruiting for the mines. Of the two homelands, the Transkei supplied the majority of migrants, increasing its complements between 1977 and

1985 from 90 000 to 135 000 workers. The Ciskei's labour force, on the other hand, declined over the same period from 14 000 to 12 000. In general, though, the Chamber found enough workers in the Eastern Cape to replace those foreign workers lost during the crisis of 1974–5. This was not a new departure, as the region had often served as a back-up supply area during times of crisis.[38]

Yet, having tapped this reservoir in their hour of need, Chamber members began to turn their attention in the late 1980s to how labour from the Eastern Cape could be limited. The issue started to weigh heavily on the Chamber for a number of reasons. One was the traditional concern with and fear of over-relying on one source area. The mines felt that drawing nearly a quarter of their labour needs from the Eastern Cape placed them in a potentially precarious position. In the 1980s, Anglo American began as well to press the Chamber to limit recruiting from the area because Xhosa-speakers from the Eastern Cape were seen to be behind the militancy of the NUM and violence on the mines.

To limit employment, the recruiting agencies first stabilised the existing labour force by placing an increasing number of workers on one-year contracts. In 1985, about 50 per cent of Transkei's labour force were on contract, and by 1988, 90 per cent.[39] Of these, 95 per cent were recruited on one-year contracts. The agencies also began to reduce the number of novices recruited in the Eastern Cape. By the late 1980s, the percentage of workers who were novices had dropped considerably below the average for the industry. Workers from the Ciskei were similarly restricted.

In response, officials of the two homeland governments urged the Chamber to maintain recruiting levels. They were critical of any attempt to limit recruiting, and worried about the impact on their states' revenue-base. When it became evident that recruiting levels might decline, homeland officials complained to the Chamber.[40] They were concerned, too, that the Chamber would alter recruiting patterns because of 'the number of Transkeians who were involved ... in unrest' and other industrial disturbances, particularly strikes.[41] A senior official from Ciskei believed that recruiting was affected by the 'attitude of Ciskeian migrant workers' who persisted in 'playing roles in industrial strife ... under the influence of militant trade unions'.[42]

The homeland officials were willing to go to considerable lengths to discourage their migrant workers from joining the NUM and participating in work stoppages and strikes. Indeed, the Ciskei government became notorious at the time for its repressive policies towards African unionism. Transkeian officials assured the Chamber that they would 'educate ... mineworkers and ... show them the benefits and consequences their behaviour has on the mines'. They promised to 'to publicise to citizens the importance of keeping their jobs in mining for their own good and for the income derived in Transkei from their earnings'.[43] In a number of official propaganda campaigns, workers were informed that they did not have to join the NUM. During the 1987 African miners' strike, workers were urged to return to work. At the same time state delegations consulted with officials from the Chamber, and asked them not to blame the strike on workers from the Eastern Cape.

It is not clear from the evidence, however, whether workers from the Eastern Cape were in fact particularly active in the strike. True, the NUM's membership base was drawn heavily from the Eastern Cape and Lesotho, but that reflected the demography of the mine labour force. Certainly, African workers from the Eastern Cape have historical roots in patterns of industrial militancy that go back to the 1920s and 1930s.[44] In the course of the development of the independent trade-union movement in the 1970s, workers from the Eastern Cape proved particularly important. NUM organisers during the early phase of unionisation were mostly educated unionists from the Eastern Cape.[45]

But whatever the truth of the matter, the result of the Chamber's policies was a reduction in workers hired from the Eastern Cape. Blamed as too political and prone to excessive absenteeism, Ciskei's labour force declined from 14 434 to 12 000 and Transkei's from 126 511 to 114 893 workers between 1984 and 1989.

The Changing Functions of the Labour Bureaucracies

The stabilisation of the mine labour force has since the early 1980s brought about a number of changes in the function of the TEBA recruiting offices. As noted before, a majority of all African workers, foreign and domestic, now hold one or other stabilisation certificate and are recruited on one-year contracts.[46] Except in cases where there is unexpected work, or when requisitions for labour are made during strikes, in general the TEBA offices forward workers already employed and on contract: 'There is no doubt that by and large our offices in the field are now primarily involved with the forwarding of labour in possession of industry certificates. Orders are seldom received and more often than not are specific as to requirements.'[47] The recruitment of novices and those without industry certificates has become a small part of the functions of the TEBA office.

As a result, the Chamber has rationalised TEBA's recruiting network, shutting down many of its offices, and retrenching staff. After 1980, TEBA closed twenty of its offices in South Africa alone.[48] TEBA offices in Ciskei were merged with those of Transkei, the Pietermaritzburg office merged with KwaZulu, and plans have been made to shut down the Bloemfontein office. In Mozambique, the twenty-one WENELA offices were reduced to four.

From the beginning of the 1980s the unemployed, those with no previous mining experience and those who failed to meet stated requirements, had difficulty getting access to jobs. Consequently, the individual TEBA office became overcrowded, faced long employment lines, and had an overworked staff. As a recruiting agency, the TEBA field office had an interest in processing as many workers as possible, and was obliged to pay attention to all applicants who put in an appearance. Workers were routinely turned away. At the Bophuthatswanan office in 1987, for example, 'large numbers of prospective workseekers continue to be turned away at all field offices daily as no employment is available to them'.[49] Similarly, at the KwaZulu office, there were 'in excess of 400 workseekers at each office, daily'.[50] Often, the workers who were turned away travelled far in search of work: 'Hundreds of workseekers are being turned away weekly and in numerous cases, particularly at Grahamstown and Fort Beaufort, men have travelled hundreds of kilometres to reach TEBA offices only to be told that there are no vacancies.'[51]

Problems of social control at the field office emerged when workers decided to squat, waiting for jobs. At the Ciskei office, for example, the staff were 'obliged to deal with thousands of desperate workseekers who do not believe that there are no vacancies'.[52] At the Transkei offices 'many workseekers sleep on the pavement outside the office in their desperation to obtain employment'.[53] TEBA's field offices in Mozambique and Lesotho reported similar problems. Workers laid siege to the offices, camping and sleeping rough in the neighbourhood, sometimes for very long periods.

During the 1987 strike, the overcrowding and 'camping' problem was made worse because many of the dismissed workers simply refused to leave the field offices without re-employment. Other workseekers believed that the strike raised their chances of employment. At the Bophuthatswanan offices, for example, a majority of the miners who were dismissed by Vaal Reefs, an Anglo American mine situated on the West Rand, and who were not re-employed 'refused to go home and have taken up residence in the streets around the office'.[54] At the time, the same office reported that 'news of the strike has raised the hope of the unemployed of consequently obtaining employment'.[55]

During normal times, two avenues were open to desperate people seeking work. As noted before, one strategy was persistence – camping at the field office until something came up; but it often ended in disappointment. At the KwaZulu office, recruiting staff reported in 1986 increasing incidents where work-seekers became 'aggressive at times'.[56] The Free State office reported that experienced workers who did not have stabilisation or leave certificates were aggrieved by the alleged injustice of TEBA's hiring practices, and pelted the cars of staff and the field office with stones.[57] The Free State was not alone in this regard. In the end, necessity drove many back home. The Ciskei TEBA manager remarked that 'these men have become despondent and can no longer afford to call at offices on a daily basis on the off-chance of being offered employment'.[58]

The other strategy open to work-seekers was to 'shoot straight'.[59] Instead of giving up and returning home, many work-seekers bypassed the TEBA office and went straight to the gold mines. In the Ciskei and Transkei it was rumoured that 'the only way to get a job is to visit individual mines'.[60] It was a source of great irritation to the field-office staff that workers would go directly to the goldfields, and that the mines would actually hire them, so undermining the recruiting system. Describing the unemployment problem in the Ciskei, the field manager explained how the 'shoot-straight' system worked: 'This situation is aggravated by the number of men who apply directly to a mine for a job. Upon acceptance a letter is forwarded to this office authorising his engagement and word of such engagements spreads rapidly causing discontent amongst those seeking employment in the field.'[61] Field managers also feared that with the lifting of the state's influx control system in 1986, workers would feel encouraged to 'travel independently to mines where they may stand a better chance of gaining employment than at a field office'.[62]

It is not clear from the evidence how widespread the practice of 'shooting straight' became. Though recruitment and employment statistics did not record the incidence of direct hiring at mine level, 'shoot-straight' cases were reported regularly from the field, suggesting that the problem was a persistent and growing one. Proceeding directly to a mine was not open to everyone, however. Resources were required for transport, and applicants had to take the risk of not being hired at the mine.

'Shooting straight' was not a new phenomenon. Alan Jeeves has documented that as late as 1916, a majority of mine workers from the

Cape were unrecruited voluntaries, making their way to Johannesburg without the assistance of a recruiter.[63] In the 1980s, as Stanley
Greenberg has indicated, the practice had far-reaching repercussions
on the systems of state control and worker recruitment which had
been built up in the intervening decades. In effect the influx control
system and the functions of the state's labour bureaucracies were
more and more subverted by workers who went straight to places of
employment in the cities.[64]

Conclusion

In the aftermath of the foreign labour crisis, the Chamber turned
towards the homelands to solve its labour needs. It did so in a context
created largely by state officials who, in pursuing apartheid goals and
by resettling Africans and by controlling their movement, turned the
homelands into vast rural slums of unemployed people. Mine recruiting thus came to take place in a flooded labour market.

As a result, the Chamber was able to pursue a policy of
heterogeneous sourcing, and mines could choose their labour from a
variety of competing sources, selecting their labour carefully. This
now made the stabilisation of the labour force possible, so that
migrant workers could return to the same mine and often to the same
job and worker-group year after year. In consequence, the task of the
mines' labour bureaucracies become one of forwarding, rather than
actively recruiting, labour. In this labour market, the more active
functions of TEBA diminished and its administrative character
changed.

Theories of the labour market for the gold mines have been concerned with the reproduction of the individual worker in the context
of the household economy, declining subsistence production, and the
gender division of labour. These theories focused attention on
homeland workers who already had jobs, and sought to explain how
renewal functions performed in the homeland lowered the costs of
their reproduction for employers. Though important, this approach
does not say very much about the character of the labour market in
the 1980s. By then the homelands had come to consist of proliferating
slums with populations having little access to an independent means
of subsistence. Jobs had become scarce; and mine jobs, previously
only taken as a last resort, were now in high demand. Moreover, the
Chamber's stabilisation policies had the effect of further limiting
employment opportunities. For the people of the homelands, access
to jobs in the 1980s became much more important than the 'renewal'

of workers already in employment.

After the 1987 strike, jobs in the mines began to decline. Not only did mine management pursue leaner hiring policies, but the introduction of technological innovations resulted in retrenchments. Between 1987 and 1989 a total of 68 000 jobs were lost. In 1990 the unfavourable gold price placed 22 mines on the marginal list, threatening the closure of some mines and shafts, and creating the possibility of further job losses. And structural problems of access to mine jobs for homeland workers worsened when gold came under increasingly unfavourable international economic circumstances, even after the outbreak in January 1991 of the Gulf War.

The scramble for jobs among people from increasingly impoverished homelands, documented in such vivid detail by Wilson and Ramphele in *Uprooting Poverty*, is the legacy apartheid leaves for state and society in the post-apartheid era. There are concerns that when the last restrictions on African mobility are finally lifted – the Group Areas and Land Acts – the jobless will pack up and move to the city, transporting the unemployed from their invisibility in the homelands to the large and growing squatter settlements around urban centres. Whether or not such demographic shifts will occur, it remains incontestable that the 'new' South Africa will have to face an unemployment problem of almost insurmountable proportions.

5
Urban Labour and Mine Housing

Introduction

In the 1970s and 1980s the mines made a number of forays into urban labour markets in South Africa. During the foreign labour crisis of the mid-1970s, for example, the Chamber opened labour offices for the first time in African townships, with the specific intention of recruiting urban workers on the Witwatersrand. During the early 1980s, Rand Mines experimented with township workers in its East Rand Proprietary Mine (ERPM), in response to the urging of local officials who pressed for greater involvement by the mines in relieving unemployment in local townships. These initiatives were shortlived, however. Chamber officials decided to close the urban recruiting offices merely two years after opening, and Rand Mines concluded the labour experiments, unenthusiastic about their future prospects.[1]

The Chamber's figures indicate that 5 184 urban workers were recruited overall in 1974; this increased sharply to 39 358 workers by 1976. The response of the workers varied from city to city. In some centres, Cape Town and Port Elizabeth for example, few workers signed up for work, and the recruiting offices were forced to cease operations fairly soon. On the other hand, workers from the townships of the Witwatersrand – where the Chamber showed the greatest interest – were extremely eager to take mine jobs, and recruiting expanded here from 1 809 to 13 007 workers between 1974 and 1976.[2] Similar expansion occurred in other cities; in East London recruitment rose from 2 196 to 11 601 workers in two years. As for the ERPM experiment in the early 1980s, management reported that close to 2 000 urban workers applied for mine jobs, a relatively large figure for a single mine.

Table 5.1. Urban recruits by city, 1973–1985.*

	1973	1975	1977	1979	1981	1983	1985
Cape Town	230	596	1 500	--	--	--	--
Port Elizabeth	543	1 740	2 948	639	--	--	--
East London	1 403	6 092	11 601	4 344	4 134	2 668	2 206
Kimberley	--	876	1 952	819	409	496	805
Bloemfontein	1 881	694	3 522	2 597	2 413	1 961	4 046
Durban	--	--	3 926	1 719	1 449	885	1 027
Pietermaritzburg	1 034	3 909	4 811	1 936	1 810	1 213	1 335
Pretoria–Witwaters-rand–Vereeniging	93	7 991	10 414	3 145	1 788	3 478	3 807

*Workers recruited at urban offices. Over 80 per cent were without rights of urban residence under influx control. Many originated from rural areas but registered for employment in cities.
Source: Jonathan Crush.

On what grounds, therefore, could it be claimed that urban recruiting was a failure? If anything, the response of urban workers was surprising, given the unattractiveness and inherent dangers of underground work, and the fact that the wage rates were pegged at largely uncompetitive migrant levels. What is striking about the experience of urban workers was that they did not stay in mine employment for long. When the opportunity arose, they left their mine jobs as quickly as possible. A recurring complaint of management was the poor retention and high attrition rate of urban workers. If anything failed, it was that the mines proved unable to keep their non-migrant labour.

Much of the problem derived from the expectation of both management and migrants that the urban workers would adjust to a migrant-based mine culture. Both migrant worker and managerial groups insisted that urban workers adopt the long-established cultural practices of the mines, and saw assimilation as a one-sided process. On average, urban workers were better educated than migrants, but because they were novices, they were placed in the most menial jobs. When it came to promotions, the urban workers were more readily passed over, because management had already decided that they were 'unsuited' for mine jobs. Urban workers also resented the fact that they had to share living quarters and food with migrants. Their answer to the problems posed by mine culture was to abandon their jobs as soon as possible.

Thus in the 1980s, when the mines came under considerable pressure to search for alternatives to migrant labour, management evidently preferred to seek ways of settling and housing existing migrant workers with their families on or near the mines rather than hiring workers from the townships. Such a route avoided the assimilation problems that past experience with urban workers had revealed. The housing initiatives, however, were only to benefit the more senior African employees in higher-income categories, and this in turn reinforced the growing class division in the African labour force. In the past, labour-repressive processes – the colour bar in particular – had limited the occupational advancement of African workers and skill differentiation among them. Despite some stratification along skill lines, African workers had tended to fall within narrow class boundaries, forming in management's eyes a relatively undifferentiated labouring mass.

In the 1970s, this pattern would begin to change. Adjustments in the application of the colour bar promoted increasing class differentiation between sections of the African labour force. More skilled, supervisory and managerial jobs became available to Africans. And yet, until the 1980s African workers all shared the same hostel accommodation: greater class differentiation at the workplace was not accompanied by diversified housing policies. When in the 1980s conflicts on the mines increasingly occurred along class lines, and a number of African 'team leaders' lost their lives in clashes between ordinary migrant workers and their supervisors, management came to believe that undifferentiated housing practices had politicised emergent class relationships. In consequence, the mine industry sought ways of taking senior African employees out of the hostels. Viewed in this way, one of the most prominent features of labour practices on the gold mines in the 1980s – the expansion of mine housing – can be seen as an answer to the problems created by class conflict between sections of African workers.

Township Labour

In a competitive labour market, urban workers in South Africa preferred to take up jobs in the manufacturing and service industries. When the jobs were available in these sectors, there never was any doubt as to their employment preferences. Historically, mine wages had been too low and conditions of service too dangerous and unattractive to draw urban workers. Instead, the mines chose to search for migrant workers from the rural areas of sub-Saharan Africa to work at low rates of pay. The improbability of attracting urban workers in a migrant-labour market was in fact generally recognised in the mining industry. In 1986, for example, a study group attached to the Chamber argued that the mines could not expect to attract and retain urban labour with 'its present migrant wage policies'.[3] However, under circumstances where the jobs were not expanding in preferred sectors, urban workers increasingly looked to mine work in spite of its disadvantages. Long periods of unemployment made work-seekers desperate, and drove them to find work wherever it could be had, low wages or not.

Once hired by the mine, urban workers were confronted with a foreign and often hostile cultural environment. As minorities in the labour force, they were expected to make all the necessary adjustments to the prevailing mine culture. They had no choice but to live with migrants in the single-sex hostels, even though some had their families resident in nearby townships. They shared the same diet arranged specifically for migrants, which included delicacies like offal; and within the all-male compounds they were subject to widespread homosexual pressures.[4] At the workplace, management expected them to be unassertive and deferential, and to observe the established racial etiquette. Many, though not all, of the line managers insisted on being addressed as *baas* (boss), in recognition of their authority and power.[5] And when the workers evidently failed to meet these expectations, they were told that they took poorly to 'mining discipline' and had 'settling in' problems.[6]

Moreover, both established migrant groups and management expected urban workers to learn *fanakalo* as a matter of course. A phenomenon deserving of greater research, *fanakalo* arguably began as an industrial creole language used as a means of communication in a multilingual and multi-ethnic setting.[7] Its origins were rooted in communicative necessity between various groups of people – white and black – who shared no base-line language. Over time, it became in practice a language of instruction and command – 'well endowed with imperatives and little else' as Charles van Onselen put it – along the chain of command between management and miners.[8] Since class and race coincided in this instance, it was used by whites to tell Africans what to do.

Urban workers were largely English-speaking, and saw no reason why they had to converse in *fanakalo*. They had more formal education than migrant workers, and were more assertive. However, because they were novices, management allocated them menial and 'labouring' work. When they complained, they were portrayed as 'cheeky' and 'aggressive', and passed over when it came to promotions. This produced a constantly reinforcing cycle of grievance. With no union or independent means of representation at the time – the mid-1970s – the urban worker's protest was to refuse to speak *fanakalo*.

The use of language to subvert the established order was developed most effectively by black Rhodesian (later, Zimbabwean) workers. As we have seen, the recruiting of Rhodesians, mainly from the townships of Salisbury and Bulawayo, was one of the Chamber's answers to the foreign labour crisis of the mid-1970s.⁹ Like their South African counterparts, the Rhodesians were more literate, articulate and politically assertive than fellow migrant workers, but found themselves placed lowest on the occupational ladder. Management was loth to promote them because of their 'arrogance', and could think of nothing better to do with them than eventually terminate their employment.

Language became the medium through which the Rhodesians displayed their status as urbanites; it signified linguistically their particular position in the industry. They insisted on speaking English to everyone, black and white, adamantly refusing to speak *fanakalo*. This served two subversive purposes. In the first place, it was an active undermining of managerial authority. Since *fanakalo* was a command language, the Rhodesians questioned the presumption that black people were simply the recipients of instructions. The use of English placed the Rhodesians on a level of equality with whites, and above those who were themselves not native English-speakers. The strategy had a familiar ring to it. To insist on the use of English meets the English-speaking South African on his or her own terms; to promote a non-South African inflection rubs the equality in. What is more, for an African to force an Afrikaner – and most of the line managers African workers encountered in the production process were Afrikaners – to converse in the language of his former colonial adversary was an act of some symbolic significance, especially after the Afrikaners had made the effort to learn *fanakalo*.¹⁰ To have an Afrikaner struggle and fumble with English was to reverse the typical situation of power experienced elsewhere in South African society. One Afrikaans-speaking manager resented the fact that the Rhodesians would 'ridicule' and 'ignore authority'; when addressed in *fanakalo* they replied in English; threatened, they would 'laugh'.¹¹

Secondly, the use of English reinforced the line drawn between urban-based and rural migrant workers. As J. K. McNamara has noted, in a situation where worker groups competed over access to jobs and promotions, ethnic and national markers came to define the lines of competition. Migrant workers saw urban workers – the Rhodesians, in particular – as threatening their access to more senior jobs, by virtue of their apparently greater sophistication, higher education, and fluency in English. Different food-tastes (the

Rhodesians refused to eat offal) and dress style (Rhodesians typically wore sunglasses and jackets) were taken as visual confirmation of the differences between worker-groups.[12] Since many of the migrants of rural origin could not follow the conversations conducted in the medium of English between the Rhodesians and whites, they felt excluded from and resentful of a presumed, but non-existent, alliance which they suspected of conspiring to frustrate their access to jobs and promotion prospects.

By 1976 the Rhodesians had developed a reputation among management for being 'arrogant' and 'troublesome'. A general manager of one mine felt that things had become so bad they 'could quite easily do without these workers'.[13] Another complained that the labour problems experienced at the time 'could be attributed to them'.[14] Tension over the Rhodesian workers became so urgent on one mine in the late 1970s that a group of migrant workers plotted to drive them off mine property.

By 1977 the argument against the further recruitment of Rhodesians was so evidently overwhelming that the Chamber took steps to phase them out of the labour network. From February 1977 only those with valid re-employment guarantees were hired. By 1979 employment had declined to 7 000 workers. When the new independent government of Zimbabwe revoked TEBA's recruiting licence in 1981, management expressed itself pleased with the departure of the black Zimbabweans, who by 1982 had all left the mines.

The ERPM Experiments

In the early 1980s, Rand Mines' ERPM opened a new shaft that would add considerably to the working life of this long-established mine, and sought to build a hostel to accommodate 6 500 additional African miners needed for work.[15] The mine was situated in the constituency of Boksburg, whose member of parliament, J. P. I. Blanché, held apartheid-style beliefs. Blanché objected to the proposed siting of the new hostel, protesting that it would be situated too close to white neighbourhoods.

The thrust of Blanché's campaign against the ERPM centred around the mine's labour mix. Like officials in the Department of Co-operation and Development, Blanché believed that the mines should hire fewer foreign and more domestic workers,[16] and urged the ERPM, historically dominated by foreign workers, to help solve the local unemployment problem by hiring workers from the African townships near to Boksburg.

It was not the first time that Blanché had criticised the labour policies of the ERPM. The issue was first raised early in the 1980s, and the mine had responded by requesting the East Rand Administration Board to forward unemployed Africans from the neighbouring Vosloorus township to the mine for surface work. 'The response was most disappointing,' a TEBA official noted. Between April 1983 and March 1984 a total of 30 workers took up employment at the mine.[17] But Blanché persisted, and Rand Mines felt compelled to take more definite steps. Although the Government Mining Engineer had already given ERPM permission to build the proposed migrant hostel, Rand Mines decided to launch an experimental project whereby urban labour from the three surrounding townships – Vosloorus, Duduza and Tsakane – would be hired.[18]

ERPM was not the best choice for the local recruiting experiment. It was an old and deep mine, known for its poor working conditions. Management recognised that ERPM was 'generally unpopular' because of the extreme depths at which mining takes place and the arduous nature of the work processes.[19] The reputation of the mine can best be illustrated by two sets of statistics. One set, compiled by J. K. McNamara for the months January–June 1976, measured the popularity of mines by the proportion of novices hired. If the proportion was high, the mine was sure to have a poor reputation, for it meant that few miners returned and many novices needed to be hired. In these terms, of all the operating mines in 1976, ERPM was the least popular.[20] Another set of statistics, recording the reportable accidents published annually by the Chamber,[21] can also be taken as a measure of a mine's reputation. In ERPM's case, it had the least favourable accident rating of the 40 operating mines in 1985 and 1986. A product of its poor reputation was the high concentration of foreign workers, who typically were more vulnerable in the labour market and therefore less reluctant to work under unfavourable conditions. Of ERPM's labour, 72 per cent was foreign, while the average for the industry was 40.

For all these reasons, members of the Chamber of Mines Research Organisation (COMRO) involved in the ERPM project made it clear to TEBA officials that the mine was a 'bad choice' for such an experiment, claiming that TEBA would not 'even get rural South African blacks' to work there.[22] A retired TEBA official living in Cape Town was asked to interview 200 'desperate' men in Crossroads, to ascertain their interest: there were 'no recruits'. However, once ERPM put the word out in the local townships that jobs were available, the mine was 'deluged by hundreds of applicants'.[23] As COMRO officials claimed, the positive response had to do with 'large scale unemployment in the townships'. They believed that a large pool of labour was available, which could be successfully tapped.[24]

Some 2 000 applications for mine jobs were made. However, ERPM only wanted to employ married workers who qualified under apartheid laws to reside legally in cities (especially the Section 10 clause of the Natives (Urban Areas) Consolidation Act of 1945). This disqualified all but 144 workers, to whom employment was offered. Of these, 9 workers turned down the jobs and 47 failed to turn up for work; the remaining 88 accepted.

The rate of attrition among the remaining sample workers proved rapid. Of the 88 workers, 29 were still in their jobs after four months and 16 after a year. Officials from COMRO blamed the workers. The men, they noted, 'did not take favourably to underground labouring work'. Interviewed as to why they had left, the workers asserted that 'wages were very little'; they felt they had been 'badly treated by superiors', Afrikaner line managers in particular holding 'negative attitudes' towards them because they were from the townships; and they claimed that African team leaders were 'prejudiced' and migrant workers as a rule 'ostracised' them.

ERPM and Rand Mines declared the urban recruiting experiment a failure. Although TEBA simply blamed the workers – they were not, an official asserted, 'interested in underground employment' – the problem lay elsewhere. The mines had made no special provision for the acculturation or integration of urban workers into a predominantly migrant milieu; and programmes had not been introduced to ease their entry into an alien sphere of employment.

The Changing Class Structure of the Labour Force

In the 1980s, the mines developed an interest in the provision of housing and accommodation alternatives directed at sections of the African labour force. Although attempts still continued to be made to recruit urban workers, some successfully, mine management now preferred to settle migrant workers in company or privately owned houses near the mines, rather than hiring workers from nearby townships. This interest in housing came from several quarters. Industrial relations and labour intellectuals attached to the mining houses had come to the conclusion that the militancy of African miners in the 1980s was partly rooted in the fact that all African workers regardless of their job status were compelled to reside in the mine hostels (though some enjoyed separate and superior quarters within hostel complexes). Indeed, the campaigns of the NUM against migrant labour and hostel accommodation highlighted management's argument that the common housing circumstance of a class-divided work-force had become a major factor behind the politicisation of industrial relations.

Historically, the African labour force in the gold mines was relatively unstratified. Although there were some differences in job content and a degree of occupational differentiation, African miners tended to work within a narrow skill band: the majority were labourers involved in menial work. On top of this, the colour bar placed a ceiling on how far African workers could advance, typically not beyond a few categories of semi-skilled work. As a result of their class homogeneity, African workers were often seen by white workers and management as an undifferentiated mass who held the same inferior status.

Except for a very small number of workers, Africans were compelled to live in compounds regardless of their particular job or income. This was as much a result of mining-house practices as it was state policy. For the mines it was evidently cheaper to accommodate migrant workers on a temporary basis in compounds. For their part, apartheid officials invoked a rule which specified that no more than 3 per cent of the labour force could be placed in family-type housing. Indeed, the migrant system enforced on the mines became the envy of apartheid planners, and was regarded as a model worthy of wider emulation. It brought workers into the cities to perform a particular task, provided them with temporary accommodation, and after completion of the job, ejected them to their place of origin.

Over time, however, the class character of the African labour force on the mines changed. In a later chapter dealing with the colour bar, I shall show how white workers bargained with management in the 1960s and 1970s over the operation of the colour bar, and in exchange for higher wages and better conditions of service ceded more and more skilled tasks to African workers. Partly as a result, the percentage of African workers classified as semi-skilled increased from roughly 22 to 32 per cent, and those skilled from 2 to 7 per cent, whereas those unskilled declined from 68 to 53 per cent, between 1960 and the 1980s.[25] It was expected that by 1990 the proportion of unskilled workers would drop to 46 per cent, that of skilled rise to 10, leaving 42 per cent in semi-skilled positions.

Increasing skill differentiation was accompanied by greater class stratification. More African workers became junior-level supervisors. By the 1980s, the most senior jobs on the mines for Africans were 'artisan aides' and 'team leaders', the latter job involving a wide range of supervisory tasks. Team leaders were the immediate supervisors of African work-gangs, and were directly responsible for a shift of work. What is more, in anticipation of the lifting of the colour bar, some mines already began in the early 1980s to train African workers for more senior work, and promoted many to supervisory jobs. By 1987, when the colour bar was about to be legally abolished, the mines employed 780 Africans as underground officials and 2 414 as surface officials.[26] These were in addition to 'team leaders' and 'artisan aides', categories of work from which many of the new officials were recruited.

In consequence, the life-chances of African workers were no longer so clearly shaped by their racial status in the labour-repressive framework. In turn, their responses to the changing labour environment came to be informed more by emergent class interests than by a common racial solidarity. During strikes and work stoppages in the 1980s, for instance, senior African employees were often reluctant to support union campaigns, although they had initially been active in the NUM. They did not necessarily share with other workers the same interest in work stoppages and other campaigns, nor did they always wish to participate in their class struggles. Partly as a result, African supervisors came to be seen by ordinary workers as having interests closer to management, and were often accused of being *impimpis* (informers).[27]

In his study of mine violence, J. K. McNamara has indicated how inter-group conflict on the mines in the 1980s became increasingly articulated along class lines. Ordinary workers resented the privileges senior African employees visibly received, particularly their superior sleeping-quarters, beds, bar facilities and food service. In several episodes, workers invaded the sections where senior African employees lived, and helped themselves to food and liquor. During 1986, when violence in the mines was particularly intense, a number of team leaders were physically assaulted, and in one particularly violent incident 8 were murdered.[28] Of the 330 workers who lost their lives in mine-related conflicts after 1974, 118 died in 1986 alone, a year characterised by intensified class conflict between African workers and management.

Management recognised that African supervisors lacked authority in the workplace because they were forced to share quarters with ordinary migrants. Their 'on-the-job' power was undermined by having to live 'in single hostels' with workers, as one mine report stressed.[29] Authority in the labour process was not reinforced by similar authority in the living environment. For the mines, the senior African workers represented a considerable investment, and were not easily replaceable. They had been put through special training programmes and held important jobs. In management's reckoning, the reproduction of the senior sections of the labour force was increasingly incompatible with traditional housing policies. In 1988 a mine report recommended that since 'the relaxation of job reservation measures and the subsequent introduction of blacks to jobs previously held by whites, we should employ blacks under stable family conditions'.[30]

Over time, hostel accommodation had itself become politicised and challenged by African workers. In a later chapter, I indicate how union members used the hostel to enforce strikes and discipline non-striking workers. Just as the hostel had been a means by which management could discipline the residents, so unions would use it to serve their own purposes. Stabilisation processes had also changed the social character of ordinary migrants, who were more and more unwilling to put up with the contrived nature of hostel life. Longer periods of employment and shorter periods of leave produced a greater commitment on their part to pursue mine-related interests. Furthermore, union membership and worker-education programmes heightened social awareness and sharpened political consciousness about possible alternative housing arrangements. By the 1980s, it was becoming more and more apparent to management that hostel administration was 'increasingly difficult to manage' and 'vulnerable' to union agitation'.[31]

Mine Accommodation in the 1980s and 1990s

It was change in state policies in the 1980s that made it possible for the mines to consider alternative forms of accommodation, particularly the introduction of home-ownership for their African employees. In the late 1970s, the state began to recognise limited forms of property rights for Africans living outside the homelands. By 1987 it had completely abandoned the system of influx controls, used historically to restrict African urbanisation by policing access to urban housing, jobs and past residence in cities. Stanley Greenberg has convincingly argued that influx control was lifted because the system had become a largely ineffective and costly instrument of regulating urbanisation.[32] Africans wishing to move permanently to cities could now in principle do so without prior access to housing or jobs, although other apartheid measures still regulated movement, such as homeland citizenship and squatting laws. Group Areas and related legislation ensured, too, that Africans confined themselves to established or new townships, separate from areas inhabited by white, coloured or Indian people.

Of particular relevance for the mines was the withdrawal of the regulation that prohibited the mines from providing family housing to more than 3 per cent of their African labour force. This regulation had been used to restrict African settlement on the mines, and historically represented a somewhat pointed response to Anglo American's request made in the 1950s to house up to 10 per cent of its African labour when the Free State goldfields were opened. The existence of the regulation became for the industry a useful defence against critics of migratory labour and mine hostels. In fact, the mines never took full advantage of what the rule allowed, settling no more than about 2 per cent of African labour in family-type housing on mine property.[33]

The responses of the mining houses to the pressures for alternatives to migrant labour within the revised state framework varied considerably.[34] Some of the mining houses, such as Gold Fields and Anglovaal, indicated little interest in abandoning hostels or developing new housing schemes or alternative labour forms. Genmin promoted the concept of an 'all-inclusive wage', which when implemented would provide workers with the resources needed to secure their own form of accommodation. The proposals of Anglo American, JCI and Rand Mines were more far-reaching, putting in place home-ownership schemes, rent and mortgage subsidies, and 'living-out' allowances, while the mines continued to hire workers from townships. Because of the different responses, a centralised policy, agreed to by all the members of the Chamber, was not possible. Housing became a mining group, and not a Chamber, concern.

A number of authors have suggested that the responses of the mining houses followed either 'reactionary' or 'progressive' corporate directions.[35] This is not a very useful distinction. Mine policies often bear very little relation to the public–political profiles the mining houses desire to project. Public profiles, moreover, change over time, a fact illustrated by Anglo American's harsh response to the 1987 miners' strike, which was arguably at variance with its liberal image. Genmin, portrayed as 'reactionary' or at least 'conservative' because of close ties to institutions of Afrikaner nationalism, became committed to ending paternalism in housing policies and ceased construction of hostels in 1990. What seems in this instance to be the crucial variable is not the public profiles of corporations. Rather, at those mining houses where industrial relations divisions which support accommodation alternatives were strong, and where the NUM was well organised, housing policies for African employees tended to be the most advanced.

Two of the mining houses – Gold Fields and Anglovaal – remained wedded to traditional forms of accommodation. Gold Fields management believed that home-ownership was not a desirable strategy, and that company-provided accommodation should remain the preferred way of housing workers. Resources were channelled into upgrading single quarters, constructing visitors' quarters, and expanding married and apartment accommodation. Gold Fields provided separate, superior quarters for its senior workers, but within the framework of company-provided accommodation.

On the other hand, Genmin pushed for the so-called 'clean' or 'all-inclusive' wage as a general answer to the housing problems of its African employees. The 'clean' wage was based on basic earnings plus the costs of mine accommodation, food, travel, and other items normally provided by the mine. Paid this wage, African workers would theoretically be free to choose their own forms of accommodation, whether they be home-ownership or rented accommodation in existing townships, site-and-service schemes, mine hostel rooms rented at market rates, or even squatter settlements close to the mines. In this way Genmin would be entirely free of responsibility for accommodation. It was planned that the new system would become fully operational by 1992, and in anticipation no further mine hostels in the old style were to be built.

The other mining houses offered a mix of systems, consisting of home-ownership, a 'living-out' allowance, and rent subsidies. JCI's 'living-out' allowance took the cost of the benefits the mines normally provided and paid them to the workers in cash, for use in renting accommodation in local townships. Anglo American and Rand Mines also offered their workers a similar allowance. In some cases, payment of the allowance had the effect of reducing hostel occupancy. However, the workers evidently still preferred to live in the hostels while they were working, returning to the townships when they were not. The development of the black taxi system made this dual form of residence possible, transporting workers to and from the hostels.

The mines have not provided figures indicating how many of their workers have used the 'living-out' allowance system. Management claimed that it was not in widespread use, and that in any case it existed alongside, instead of serving as an alternative to, the hostel system. The mines also had no control over how the 'living-out' allowance was to be spent. Many workers apparently used the additional resources to construct squatter dwellings around the mine where they were employed, instead of renting accommodation in townships. Squatter settlements have emerged closer to a number of mines, including those administered by JCI and Genmin.

The thrust of new accommodation policies centred around private home-ownership. JCI provided surety for mortgages, and assisted with the provision of loans for workers who wanted to buy land, develop site-and-service schemes, or build their own homes. Rand Mines encouraged home-ownership by underwriting mortgages. Anglo American similarly offered a subsidised mortgage with a reduced deposit and low, fixed interest rates amortised over 20 years. Extensions in existing townships in the Orange Free State (Thabong, near Welkom, and Kutlwanong, near Odendaalsrus) and in the Transvaal (Kanana, near Klerksdorp) were set aside specifically to provide homes for African mine employees. The company town of Wedela, created in the late 1970s to service Anglo American's Western Deep and Elandsrand mines, was also expanded to house more workers.

The impact of home-ownership schemes on housing practices has been limited. Four years after it was first introduced, Rand Mines noted its disappointment with the progress of the home-ownership scheme. Increased interest rates in the late 1980s had had a dampening effect on the market, and Rand Mines was forced to subsidise mortgage repayments to prevent foreclosure among existing mortgage-holders. Anglo American indicated that by 1988, 656 of its African employees (or 0,37 per cent of the total labour force) owned homes under the scheme, and another 392 had signed up; by the end of 1990, the figure had grown to 3 423.[36]

Only senior African employees have been in the financial position to take advantage of the home-ownership schemes, and then they often struggled. For most workers in lower job grades – where the bulk of the African labour force was concentrated – the average available house was out of financial reach. Management noted that of the proportion of workers thought likely to afford private housing, only 30 per cent in practice could. As a result, the participants in home-ownership schemes were heavily concentrated in the upper-skilled and supervisory categories of work. As of 1989 not a single worker in the lower job grades had yet been able to participate in the scheme.[37]

The expansion of home-ownership was also restricted by the fact that a majority of skilled and supervisory workers were in principle excluded because they were foreign. Analyses of skill structures and nationality have shown that because of their long presence in the industry, foreign workers – from Lesotho and Mozambique in particular – have come to dominate skilled and supervisory work.[38] For example, of the 20 280 higher-skilled supervisory and technical positions on Anglo American mines, 60 per cent were foreign. While the South African state has made it clear that workers from foreign states must remain migrant workers, the labour-supplying states in turn, dependent on wage remittances and other administrative revenues, have insisted that their workers remain excluded from home-ownership schemes. These workers were thus prohibited from settling in family housing units and had to remain in the hostels.

The expansion of home-ownership has also been limited by the monetarist policies of the state. The mines expected the state to carry the financial burden involved in the provision of infrastructure and services to new townships and extensions. Anglo American estimated that its programme would require over R500 (US$200) million between 1988 and 1992 for roads, water, electricity, sewerage, schools, clinics and recreation facilities.[39] Unable to provide the same services effectively to existing townships, and immersed in an increasingly severe fiscal crisis, state officials turned down Anglo American's request for the financing of infrastructure and services. During the late 1980s, negotiations between state officials and corporate management were still continuing.

Finally, it remained unclear as to whether workers themselves were willing to abandon their migrant roots. Although the results of surveys commissioned by the mines indicated an overwhelming desire on the part of married workers to move permanently to the mines, a significant section of the migrant labour force was ambivalent.[40] For decades the migrant labour system has oriented workers towards improving their rural economic base. In many areas, rights of access to land and other resources were contingent on continued migrancy. Moreover, for every African migrant employed in the mines, four to five persons in wider kin networks depended directly or indirectly on wage remittances.[41] Thus although some workers had the opportunity to move permanently with their immediate families to the mines, many were hesitant because of the social and economic consequences.[42]

The attempts of the mines to provide alternatives to migrant labour, and specifically to promote home-ownership among senior African employees as a way of generating stability and protecting the mines' investments in labour and skills, have therefore been of limited success. There was no evidence that home-ownership promoted conservative and less militant attitudes among senior employees, or that the class conflict between sections of the African labour force diminished. The NUM and other trade unions have been urged to pay specific attention to housing issues, in order to counteract management's efforts to use mine accommodation as an instrument of regulating industrial relations.[43] The promotion of home-ownership and other alternatives should be seen in the light of the proposition that trade-union militancy would arguably decline when the rights and responsibilities of private property and ownership were internalised by employees moving up the occupational structure.

Conclusion

This chapter has examined two related processes regarding the mines' search for alternatives to migrant labour. The first was that urban workers developed an interest in mine jobs during periods of economic decline in city-based manufacturing and service industries, but that the mines had difficulty retaining those workers employed. The contradictory position of urban workers in the occupational and social structures of the mine and the difficulties which aspects of mine culture created, made the assimilation and integration of these workers into a migrant world problematic. The experiences of township workers from South Africa and Rhodesia, as well as the workers involved in the Rand Mines' ERPM labour experiments, illustrated these processes.

Secondly, as a result primarily of changes in the class structure of the African labour force, the mines developed an interest in alternative forms of accommodation – with a focus on home-ownership – for their workers. For a variety of reasons, the commitment proved very uneven. Although the housing alternatives were in principle available to all workers, only the most senior African employees could afford them. As a result of the state's monetarist policies, the prohibition against foreign workers acquiring homes or settling permanently near the mines, and the ambivalence of some migrant workers towards full proletarianisation, the growth of a settled African labour force has been limited and slow.

The mines expect that proletarianisation and settlement processes will increase considerably in the 1990s. TEBA believes that about 20 per cent of the African labour force will be permanently settled on or near the mines by the end of the 1990s.[44] Most of the settled workers will be skilled and in supervisory jobs and, because of state immigration policies, will originate from within South Africa. Over time, these workers will change the dominance foreign workers have had over access to skilled jobs. The other component of the labour force will remain migrant. The labour-supplying states and societies of southern Africa continue to rely on migrancy for jobs and revenues, and therefore support and encourage patterns of oscillating migration. The dependence on mine wages in rural southern Africa will also continue to create a demand for mine jobs. Over time, foreign workers will hold the lesser-skilled and unstable jobs, and will be more vulnerable in the labour market. Mozambican state officials already complain that their workers have been downgraded more and more to the status of 'allottable labour', an industry term for workers hustled in and out of relatively unstable jobs.[45] Also, although many workers from the homelands have interests in moving to the city, not all want to sever ties permanently with the homelands.

In the 1990s the industry's labour system will become more diversified and stratified. On one side, more senior African employees will be fully proletarianised and stabilised, and be resident with their families on or near mine property. On the other, the majority of workers will remain locked into the migrant labour system. The organisation of the mines' labour supply in the 1980s and 1990s, the subject of the first part of this book, has come to resemble the vision of TEBA's general manager in the mid-1970s, Tony Fleischer, who at the time argued: 'If we regard the total mine labour force as a pyramid, the peak ... would represent those holding key supervisory and production jobs. These jobs entail costly training which is of value both to the mines and the man trained. Therefore, the top of the pyramid should be stabilised and, where possible, the workers should be South African blacks living on or near the mine with their families. If, say, 10 per cent of the total labour force were stabilised we would look to this group ... for the stability that is required for smooth industrial relations ... the remaining trapezium is represented by the bulk of the labour force which will remain migratory for some years to come.'[46]

PART TWO
The Ascendance
of African Workers

When African Workers Became Unionised

The Ascendance of African Workers

At the end of 1982 the Chamber of Mines signed a recognition agreement with the National Union of Mineworkers, permitting – for the first time in a hundred years of gold-mining – the unionisation of its large African labour force. It was, as Steven Friedman recorded, a milestone in labour history.[1] Just four years before, in 1978, Chamber members had told the Riekert Commission that black mine workers were 'simple, unsophisticated tribal Africans',[2] unfit for trade unions.[3] Within a few years the impact of economic and political processes had changed corporate attitudes and revised the approach of the Chamber towards the unionisation of the African labour force on the mines.

For a variety of reasons, the state came to abandon in the late 1970s the paternalistic works and liaison committees that had long governed industrial relations in South Africa in terms of the repressive Native Labour Regulation Act of 1911, as amended. Under this system, African workers were consulted about their grievances, but lacked independent representation by means of a trade union. The withdrawal of the state in the 1980s from a direct presence in industrial relations – first signalled by the Wiehahn Commission's recommendation that trade unions be established for all – effectively created the space for the Chamber to construct a collective-bargaining system for the unorganised African worker. But the Chamber soon found itself internally divided about the way in which African workers should be organised and incorporated into a new framework. Some of the mining houses wanted full incorporation of African workers, while others were interested in extending union rights only to an elite. Torn by internal dissension, the Chamber became incapable of formulating any coherent strategy in response.

And by the time the Chamber finally drafted its guidelines, political processes had already made full African incorporation ineluctable.

Once recognised and allowed to operate in the mines and mine hostels, the NUM developed rapidly, becoming the fastest-growing and most powerful union in South Africa. At the end of 1982 the union had 12 000 signed-up members, and by 1988 this had increased to 360 000.[4] In a labour force 500 000 strong, the union had come to represent 60 per cent of African workers within five years. As Roger Southall noted, the rapid unionisation of the labour force belied previously held academic and corporate notions that migrants were not interested in unions.[5] On the contrary, the growing stabilisation of the labour force, the increasing concentration of domestic as against foreign workers, and the greater reliance on the mine wage rather than rural production worked to make African migrants eager to join unions. What is more, the lenient recognition criteria adopted by some of the mining houses (especially Anglo American) and the initial access the NUM enjoyed to the mine hostels all facilitated the recruiting of new members.

The NUM pursued both mine-specific as well as broader societal goals. NUM leaders wanted to see improvements in the conditions of mine work, particularly with regard to health, safety and renumeration levels. They pressed for the rights of workers against unfair and arbitrary managerial treatment. They opposed the colour bar in employment and the privileged position of white miners, and criticised the migrant labour system and mine compounds for their inhumane and undesirable social consequences. At the same time, the NUM supported sanctions and disinvestment as part of a broader anti-apartheid labour alliance within the Congress of South African Trade Unions (COSATU). In 1987 the NUM adopted the political programme contained in the Freedom Charter, the guiding document of the African National Congress, and elected Nelson Mandela as its honorary life president.

During the 1980s the union made a number of gains in achieving its goals, and through its activism improved some aspects of the working-environment of its members. The union gave health and safety issues a high profile, and contributed in a number of ways to improving the procedures by which safety standards were monitored and enforced.[6] It tried to shape labour legislation and turned towards the Industrial Court for protection against mass dismissals. NUM leaders participated in the political processes by which the colour bar was abolished, while in the workplace the power of white miners and mine management was challenged. Viewed more broadly, the NUM became central to the transformation of a paternalistic system of industrial relations into a modern bureaucratic one, where the conditions of service were now negotiated and not determined arbitrarily by the Chamber.

In a number of areas, however, the capacity of the NUM to achieve some of its more central goals was limited by the fact that it was a union of migrant workers operating in a sub-continental migrant labour market. The nature of this market and the policies of the labour-supplying states both placed constraints on the union's emergent power. This was particularly the case when it came to wage-bargaining. When the union took out its members on strike, replacement labour was easily found, and the power of the strike action undermined. As a result, the NUM had great difficulty increasing the wages of its members significantly. Similarly, in other areas, such as the demand to abolish migrancy and mine hostels, little headway was made.

Over time it became apparent that class struggles in the sphere of production were much too limited a framework within which the NUM could pursue its objectives. Confined to production, the union found its power tempered by a highly unfavourable labour market. Consequently it turned to a wider political framework to pursue its goals. In the 1980s the NUM would adopt a political programme, support the nationalisation of the mining industry, and underwrite the ANC's quasi-socialist accumulation policies. It would actively participate in the struggle to end apartheid and white monopoly over state power.[7] This politicisation of the NUM, as I shall argue, followed directly from its weak position in the narrow collective-bargaining framework that was created after Wiehahn.

A Divided Chamber

Christopher Pycroft and Barry Munslow have argued that the Chamber's response to the Wiehahn Commission was part of a coherent and broad-ranging strategy developed by dominant classes and the state in South Africa to 'restructure' the industrial relations system of the mines. They suggest that the Chamber reacted to the accumulation problems of the 1970s by introducing a modified system of class domination. In this apparently new approach, the stabilisation and proposed unionisation of the more skilled section of the African labour force formed key aspects. The present chapter, however, contests this interpretation and the larger literature on 'restructuring' on which it is based.[8] It will argue that the NUM emerged because of unresolved conflict between the mining houses in the Chamber, and not because of a coherent and premeditated strategy of industrial 'restructuring'.

In general, there were a number of reasons why the Chamber developed a growing interest in trade unions for African workers in the 1970s. The established white mine unions did not unanimously reject African unions, and the South African Boilermakers' Society in particular pressed the case for African unionisation with the Chamber. Furthermore, the larger mining houses were linked within financial corporations to non-mining enterprises where unionisation was successfully proceeding, and this experience served as an example the mining industry could follow. Then again, the real wages of African workers had increased considerably in the 1970s, and the industry wanted workers to make corresponding advances in productivity. On this basis the Chamber sought to end the colour bar, and grant African workers a more direct say in matters regarding conditions of service. Trade unionism for Africans was consistent with all these concerns.

What drove the Chamber most forcibly to reconsider the paternalistic system of industrial relations, however, was the absence of an effective means of communication between African workers and management, and the consequent inability of workers to express their grievances through channels they regarded as legitimate and meaningful. The lack of an effective voice was illustrated most dramatically by the level of violence the mines experienced in the 1970s. J. K. McNamara has recorded 141 episodes of conflict between workers and management and between worker and worker during the period 1974–80, which resulted in the deaths of 330 Africans and injury to 2 100. This violence, he argues, was generated by pressures in the changing labour market, and the absence of trade-union representation.[9] Management were ignorant of workers' underlying grievances, nor did they have means of ascertaining them.[10] Making the point somewhat humorously, Steven Friedman wrote that workers at Anglo's President Steyn mine rioted in 1978 and caused a million rands' damage. When asked what provoked them, the workers replied that 'they didn't like their dinner'.[11]

It was the Wiehahn Commission that provided the Chamber with an opportunity to construct a new system of industrial relations. Under the old system, direct control by state officials over African workers had served to politicise production relations, so that protest by African workers about work-related problems automatically became a grievance as well against state intervention. Through the Wiehahn Commission, reformist officials in the Labour and Manpower ministries of the state sought to depoliticise production relations, and to minimise the state's presence in what increasingly was seen as the 'private' sphere of accumulation.[12]

However, within the Chamber, division emerged about the proposed character of the new industrial relations system. Among other items, the contentious issue had to do with trade unions for Africans. Two of the mining houses – Anglo American and JCI – believed that all African workers on the gold mines should have the right to join trade unions, expecting at the same time that very few would actually do so. The other four mining houses – Gencor, Gold Fields, Anglovaal and Rand Mines – wanted to limit trade unions by restricting membership to permanent, non-casual workers – a minority of the labour force. Because each mining house, no matter how small or large, had equal representation in the Chamber, and decision-making always worked on a consensus principle, those houses pressing for far-reaching changes in the labour framework began to view the Chamber as a liability, incapable of initiating significant changes.

Anglo American, the single most important sponsor initially of the NUM, recognised early on that the Chamber was a poor vehicle for effective change. Dennis Etheredge, head of the gold and uranium division, noted in 1977 his dissatisfaction with the Chamber's initial draft response to Wiehahn.[13] Zach de Beer, an Anglo executive member concerned with labour matters, was also discouraged with the work of the Chamber committee involved in responding to Wiehahn, noting that it was 'unlikely that the Chamber is going to contribute a satisfactory document for this purpose'. He added that the problem was a result of 'the usual division and disruption' in the Chamber.[14] Anglo was not impressed by the Chamber committee's documents on industrial relations issues, because these did not attempt boldly to anticipate the future, and because they did not represent strongly enough Anglo American's emergent philosophy.

Not wishing to leave the fate of the new industrial relations system (which included the critical question of African trade unions) in the hands of the Chamber, Anglo American proceeded in 1977 to produce its own evidence.[15] Accordingly, 'the preparation of evidence for the commission will be a major task if we are to take advantage of the opportunity to have some of the major constraints to industrial relations progress removed'.[16] Anglo American wanted its major interests to become the interests of the Chamber, and its evidence to Wiehahn the Chamber's as well: 'Anglo American must agree on its primary interests in relation to the Wiehahn Commission, and must be prepared to work for the inclusion of those primary interests in the submissions of the Chamber, perhaps even at the expense or the sacrifice of secondary interests in this area.'[17] Though Anglo American executives still pushed for a consensus, they were not willing to 'get bogged down' by the Chamber, and kept open the option 'to go to the commission directly'.[18]

In their submission to Wiehahn made in 1978, therefore, Anglo American as well as JCI argued that all African miners, domestic or foreign, career or casual, should enjoy the right to free association and to join trade unions. They stipulated, though, that foreign workers should not hold leadership positions in the union, and insisted that the state control registration procedures strictly.[19] However, Anglo American did not wish to be seen to be promoting African trade unions, and argued that the process of unionisation should first begin on a mine or plant level. Indeed, a member of senior management claimed 'that the majority of the black workers have neither the capacity nor the desire to form unions'.[20] Though it supported the emergence of a broad-based, non-racial industrial union, Anglo believed 'that no effective union representing any significant number or proportion of black workers which AAC might feel obliged to recognise will emerge in the mining industry in the near future'.[21] Having made the concession in principle, they went on mistakenly to believe that a viable trade union representing African workers would not develop.

For their part, Gold Fields, Anglovaal, Rand Mines and Gencor insisted in their joint submission to Wiehahn that a distinction be made between 'permanent industrialised black workers' and 'casual, relatively unskilled migrant workers'.[22] Only the so-called permanent African workers could, in the view of these houses, have the right to join trade unions. They further proposed that a new union would first have to register with the state before any collective bargaining could proceed. They maintained that the time had 'not yet arrived for non-industrialised casual workers ... to establish labour unions'.[23] The workers had to be satisfied with 'mine-level organisations' for 'joint consultation'.

In the first of its six reports, the Wiehahn Commission recommended in 1979 that African workers be permitted to unionise in manufacturing and service industries. Having established the principle of independent worker representation in industry, the commission would have been inconsistent to refuse it for mining. In the face of a divided Chamber, Wiehahn made the same recommendation for mining in 1981. In this way, therefore, change came in the broader political framework within which the Chamber would have to construct a new industrial relations framework, based now on the inevitability of African unionisation. As its first response the Chamber produced in 1980 a set of guidelines intended to frustrate the process. These specified that the Chamber would only deal with registered unions, denied recruiters access to mine compounds, and refused to make stop-order (check-off) facilities available. Because the guidelines tried to represent the average and most conservative opinion of a divided Chamber, they lacked in fact the full support of all the mining houses.

Elsewhere in industry, management had already abandoned the stipulation that unions required to be registered with the state before bargaining. Moreover, in instances where an attempt was made to enforce a distinction between 'permanent' and 'casual' workers, management had run into major administrative difficulties. Aware of these trends and in line with its own initial recommendations, Anglo American's representative in the Chamber argued that to restrict trade-union rights to some unions was in itself a politicising act, and could provoke the labour unrest that the industrial reforms were supposed to prevent. The most promising union on the mines, the NUM, would neither register nor restrict recruiting to so-called 'permanent' workers. By 1982, the Chamber had shifted in its opinion towards Anglo American's position. In doing so, the distinction between casual and permanent workers was abandoned, as was the registration requirement.

The Growth of a Migrant Union

As a token of its change of heart, the Chamber in 1982 granted access to a number of unions which wanted to recruit among African workers on the mines. However, none of these unions had much success. The Black Mineworkers' Union, the first union to recruit in the mines, went defunct. Other unions, such as the South African Mine Workers' Union, the Allied Workers' Union and the Federated Mining Union, made limited headway. But once the NUM began to recruit in 1982 it soon scored some visible successes, and several of the competing unions backed away.

The NUM was started by the black consciousness-oriented Council of Unions of South Africa, as a response to what the CUSA executive believed to be its obligation in the face of 'persistent worker requests'. In 1982, Piroshaw Camay, general secretary of CUSA, came away from an Anglo American seminar hosted by Harry Oppenheimer at the Human Sciences Research Council in Pretoria, with the conviction that there was space in the mining industry to create a union for African migrants. Later in the year, at CUSA's annual congress, the NUM was founded, and Cyril Ramaphosa, a lawyer employed by CUSA's legal department, was asked to serve as the first general secretary. Six months later, in 1983, the NUM signed a recognition agreement with the Chamber, which was followed by shaft steward and shop steward agreements.

The NUM thereafter grew rapidly. Signed-up membership reached 360 000 by 1988, and official membership, as enumerated by the mining houses on the basis of membership fees, 175 000. As Jonathan Crush observed, it was a spectacular growth pattern, with virtually no contemporary parallels in the industrialised and semi-industrialised world.[24] A number of predisposing factors help explain the NUM's growth. Certainly, African miners were extremely eager to participate in trade unions. Studies undertaken by the Human Resources Laboratories of the Chamber's Research Organisation during the late 1970s and early 1980s indicated a high level of grievance among African miners about their conditions of service, and a burning consciousness of their exploitation.[25] With the stabilisation of the African labour force and the shift towards domestic labour in the late 1970s, worker-groups emerged which now had regular employment in mining, and therefore could potentially become stable constituencies for an emergent union.

But all this does not explain why the NUM was so successful compared with previous recruiting efforts. Part of the reason had to do with Anglo American's initial sponsorship. Among senior management the NUM was seen to approximate Anglo American's ideal of a non-racial industrial union; moreover, the NUM's recruiting competence was widely admired. Anglo American initially adopted a non-obstructionist stance towards the NUM, granting access to mine hostels. This was critical, for the very compactness of living arrangements made hostel residents a captive constituency. Anglo also provided office and other union-related facilities on a mine level. As a result, the NUM established its presence first in Anglo American mines. Since the Corporation employed about 40 per cent of the total mine labour force, the NUM made great gains within the industry as a whole.[26] Though it extended its presence to other mines in the late 1980s, it remained essentially an Anglo American-based union.

Table 6.1. *NUM membership on selected gold mines.**

	Total employed	Recognised categories	Non-recognised categories	Percentage unionised
Anglo American (at June 1988)				
Elandsrand	8 321	5 498		66,07
Freddies	12 512	3 751		29,97
FS Geduld	14 727	9 079		61,64
Western Holdings	17 400	11 505		66,12
Pres. Brand	18 746	14 260		76,07
Pres. Steyn	18 483	10 631		57,52
Saaiplaas	13 498	7 900		58,53
Vaal Reefs	38 826	18 006		46,38
Western Deep	23 233	8 384		36,09
	165 746	89 014		53,71
Gold Fields (at June 1988)				
Deelkraal	6 472	2 387	857	50,12
Doornfontein	9 456	1 643	2 713	46,07
Libanon	7 945	664	2 523	40,11
E. Driefontein	13 032	500	5 185	43,62
W. Driefontein	15 773	–	–	–
Kloof	16 476	–	–	–
Venterspost	7 842	–	–	–
	76 996	5 194	11 278	21,39
Rand Mines (at June 1988)				
Blyvooruitzicht	11 206	167	839	08,98
Durban Deep	10 062	–	–	–
ERPM	18 284	–	–	–
Harmony	31 593	–	–	–
	71 145	167	839	01,41
Gencor (at July 1987)				
St Helena	10 824	4 724		38,00
Kinross	8 975	4 961		52,70
Winkelhaak	10 330	2 677		45,20
Leslie	5 541	3 047		55,60
Bracken	4 131	2 753		68,18
Grootvlei	6 349	5 431		75,23
Marievale	1 202	1 125		82,84
Stilfontein	11 955	4 352		47,40
	59 307	29 070		49,02
Grand Total	373 194	123 445		33,08

*All mines, except Randfontein, Western Areas (JCI); Hartebeestfontein, Loraine (Anglovaal).

Sources: Hirschson (1988); figures for Gencor mines supplied by Gencor (1987).

The NUM's initial successes were rooted in a strategy of recruiting, first of all, more senior African workers.[27] These included clerks, personnel assistants and team leaders. As senior African employees were not as numerous as ordinary workers, it was easier to obtain majority representation and official recognition in their job categories. After the initial phase of recruiting, the NUM expanded its membership base to include more and more workers from semi-skilled and unskilled categories. According to Jonathan Crush, some 70 per cent of the membership were drawn from the Chamber's job categories one to four, which were the unskilled and semi-skilled grades.[28]

In establishing itself, the NUM was supported by an extensive network of domestic and international trade unions. CUSA allocated financial and personnel resources to help NUM get off the ground, and provided a number of full-time organisers.[29] CUSA and the International Committee of Free Trade Unions also supplied additional expertise and funds, and union leaders were sent overseas for training.

CUSA's sponsorship of the NUM left a number of legacies. One was the belief that there was a pressing need to develop black leadership in trade unions, to which the NUM adhered even after leaving CUSA to join the non-racial Congress of South African Trade Unions (COSATU). CUSA unions tended to emphasise the importance of community ties and politics, but neglected shop-floor organisation, whereas the reverse was largely true for those unions affiliated at the time to the Federation of South African Trade Unions (FOSATU), COSATU's predecessor. Consequently, the NUM was unable to draw on a tradition of shop-floor organisation in the extremely difficult physical and social environment it laboured in while building enduring trade-union structures in the 1980s.

The Rights of African Workers

Eddie Webster has observed that the NUM developed 'two faces' in response to the specific character of South African mining. On the one hand, the union sought to promote and entrench the rights of African workers in production, civil society and the state. On the other, it fought for improvements in the conditions of service of black workers, focusing especially on those conditions that affected wages and benefits.[30] For the NUM, working within a society where African people were denied ordinary civil and citizenship entitlements, the emphasis on the right to be protected against abuse and arbitrary treatment was as important as the more narrow, though just as necessary, pursuit of improvements in wages and conditions of employment.

One major area of focus for the NUM became the safety of mine work and the occupational health of African miners. Underground mining in South Africa has been, and remains, notoriously dangerous. In the most comprehensive overview of safety and health issues in the mines to appear to date, Jean Leger has determined that 60 000 miners have lost their lives and over a million have been injured in mine accidents since the turn of the century. In the 1980s, 600–800 workers have died annually as a result of rockbursts or rockfall-related accidents. Leger noted that while there has been an overall decline in fatality rates in the twentieth century, few significant improvements in the performance of mine safety systems have been introduced since the 1940s.[31]

Before the advent of the NUM, management regarded safety as falling entirely within their sphere of decision-making and control. After the disasters of the 1980s, particularly those at the Hlobane colliery and Kinross gold mine, the NUM began to challenge management's control over safety, and asserted the right of workers to better protection. When an explosion at the Hlobane colliery killed 68 workers in September 1983, the NUM blamed management, claiming that the enforcement of safety standards was not as it should have been. In a sympathy action, the union brought out 30 000 African workers in a work stoppage lasting half an hour. At the subsequent inquiry required by law, the NUM obtained international experts in mine safety to represent the deceased miners. It was the first time in history that African workers had been represented by an independent legal team. At the inquiry members of management were extensively cross-examined about safety standards. According to Leger, the legal proceedings revealed 'widespread neglect' of proper procedures, and information about mine safety usually sequestered by management was made public as a result.

In 1986 a disaster at Gencor's Kinross gold mine resulted in the death of 177 miners and the hospitalisation of 234 miners for toxic poisoning. In reaction, the NUM claimed that the disaster, which came about when a leaking acetylene cylinder ignited the polyurethane foam-lining of a mine-tunnel wall, was no mere accident, but had been caused by poor enforcement of safety standards. (Polyurethane was in fact a well-known safety hazard, whose use in underground mining had been banned in Britain and elsewhere.) The NUM called for a day of mourning on 1 October 1986. This was observed by an estimated 250 000–275 000 African workers in the largest industrial action ever undertaken by African workers until then. Although the NUM tried to get international safety experts and their own inspection teams to the accident site, Gencor management apparently frustrated these attempts. At the subsequent inquiry, a number of white miners were found guilty of criminal neglect and fined for their role in causing the disaster.

As these cases illustrate, the NUM used existing but largely unenforced legal provisions to highlight the abuses and shortcomings of mine safety systems. At every accident-related inquiry held after its foundation, the union saw to it that its lawyers would be present, arguing for the rights of workers. As a result, information about safety systems and production became more widely available. The NUM also began to press for safety provisions in collective-bargaining agreements concluded with the Chamber, and in 1986 a 'security of income' clause was added which made provision for injured and disabled workers to claim compensation if they were demoted, an improved accident-leave scheme, and a promise that an attempt would be made to find alternative employment for disabled workers. Similar mine safety agreements were signed with a number of individual mines.

At the same time the NUM wanted to go a great deal further. It sought to enter into more extensive safety agreements with the mines and the Chamber, and to appoint safety stewards elected by workers to monitor safety standards independently. Requests were also made for unencumbered access to and inspection of sites as soon after accidents had occurred as was feasible under the circumstances. In other areas, such as workers' compensation, the re-employment of disabled workers, and protection against occupational diseases, the union pushed for the implementation of improved measures.

Managerial responses, however, tended to be obstructionist when it came to mine safety systems. Management strove to keep information about mine safety away from public scrutiny, and sought for answers to safety problems in new mine technologies. Even in the late 1980s white miners still earned large bonuses when African workers took risks on their behalf, a practice that encouraged negligence.[32] After Kinross, the Government Mining Engineer issued regulations compelling the mines to introduce better alarm systems and safety devices. At the same time, state officials proposed that the mine-safety inspectorate be privatised: this would effectively allow the industry to enforce its own safety procedures. It was a proposal that the NUM, vitally concerned about the safety of its members, was unlikely to countenance.

The other important area of concern to the NUM was management's use of mass dismissals to break strikes or work stoppages, undermine worker-led boycotts and subvert emergent union structures. Historically, the typical disciplinary response of management to worker-led industrial disturbances was to fire workers *en masse*. In his studies, Marcel Golding has argued that mass dismissals retarded the process of unionisation in the 1980s. At Anglo's Vaal Reefs and Anglovaal's Hartebeestfontein, for example, 17 000 African miners were fired *en masse* after three months of boycotting the concessions stores on the mines, work stoppages, and working to rule for better wages and the fairer distribution of work between black and white. The mass dismissal destroyed the 'entire South Division' of the union and smashed the 'shaft stewards committee', Golding asserted.[33]

As a part of the Wiehahn reforms of the late 1970s, an Industrial Court was established with the power to adjudicate practices that in its opinion could be considered 'unfair', including the use of mass dismissals.[34] The court held out to successful petitioners the prospect of job security, interim relief, and reinstatement. Not only were dismissed workers given the chance of a hearing and an opportunity to put their case, but lawful dismissals under common law could be considered by the court an unfair labour practice. Moreover, victimised workers had a chance of reinstatement. The Industrial Court offered, therefore, a measure of protection against unfair dismissal and victimisation, in a labour market which reinforced the disciplinary powers of management.

In a number of instances in the 1980s, the NUM successfully petitioned the Industrial Court to rehire dismissed workers. In a landmark case heard in 1985, the court ordered the reinstatement of dismissed workers at Gencor's Marievale mine. After the court had found that the dismissals were unfair, the NUM and Gencor reached an agreement to reinstate 413 of the 1 000 workers fired.[35] Again, in 1986 the court ordered the temporary reinstatement of David Theko, chairman of the shaft stewards' committee at Gold Fields' Kloof mine. It also found that Gold Fields failed to give employees an opportunity to put their case at disciplinary hearings, and criticised the mine for the procedures it had used during the dispute.[36]

But in other cases, the use of the Industrial Court proved a limited instrument for checking the arbitrary powers of management.[37] In 1984, Anglo American fired 14 000 workers at the Vaal Reefs mine, and Anglovaal dismissed 2 400 workers at Hartebeestfontein. Although Anglo American and the NUM reached an agreement that the dismissed workers would be first of all those rehired, no worker was formally reinstated. In 1986, Gencor dismissed 20 000 workers at the Impala platinum mine in Bophuthatswana; police watched as the miners were bused off mine property.[38] During 1987, the Industrial Court refused to reinstate workers at the Hartebeestfontein mine because explosives had been found in the workers' rooms.

Despite these defeats, employers still believed that the Industrial Court was too concerned with the rights of workers. They pressed parliament to limit the powers of the court, and to define the meaning of an unfair labour practice more narrowly. For its part, the NUM found that the time and costs involved with cases increasingly burdensome, contrary to the court's original purpose of being expeditious and inexpensive. In 1987 Ramaphosa expressed his criticism of the fact that the Supreme Court occasionally reversed Industrial Court judgments, and that the Minister of Manpower was arbitrary in his appointment of conciliation boards.[39] Thus while the NUM at times obtained relief from the Industrial Court, it found itself more often powerless to act against mass dismissals. In defending workers' rights to job security, as in protecting their health and well-being, the NUM faced an uphill battle.

Limits of Collective Bargaining

The union's credibility among its members depended to a large degree on its success in negotiating wage increases with the Chamber. At first, the leaders had to acquire the ability to use the collective-bargaining framework to their advantage. At the first round of bargaining with the Chamber in 1983, the NUM demanded a 40 per cent wage increase across the board, to cover cost of living increases and to 'compensate them for the years when they had no say in wages'.[40] The Chamber, not known for making public statements during wage-bargaining, publicly declared the demand to be 'totally unrealistic', and made a 14 per cent counter-offer, giving the NUM one week to make up its mind. When access to hostels and halls was refused, the NUM found itself unable to consult its membership. An inexperienced union leadership fell into line, believing that it genuinely lacked time to consult with its members, and accepted 14 per cent. When members eventually were consulted, however, they rejected the offer. The union had been both outmanoeuvred and bullied by the Chamber.

Between 1984 and 1986 union leaders began to push at the boundaries of the collective-bargaining framework, and made cautious use of limited strikes. The NUM declared a number of disputes in this period and took the Chamber to a conciliation board hearing, following the requisite steps before a fully legal strike can be called. In 1984 a strike was called, but a settlement was soon reached after Anglo American made a separate offer which the union found acceptable. In 1985 strikes were called at Gencor and Gold Fields mines, but not at Anglo American, because a separate settlement was once again reached. Overall during this period, the NUM obtained wage increases greater than those the Chamber initially offered, and in this way enhanced its credibility among members and other workers.

By 1986 considerable pressure had now accumulated for a major assault on existing wage levels. The trade-union federation COSATU had launched its 'living wage' campaign in 1986, and demanded that enterprises begin to pay higher wages to all workers. On the basis of the 'living wage' COSATU was able to mount a campaign that would strengthen its relationship with community organisations and extra-parliamentary political movements, for all were concerned that wages should be high enough for families to live a decent social life. For miners, whom the industry historically paid poor wages, and to whom it justified them on the grounds that they were intended to support the worker and not his family, the concept of a living wage had great appeal.

Since its establishment, the NUM had developed into an open industrial union, with the majority of its members working as unskilled and semi-skilled operatives. To improve the wage rates of its entire membership, therefore, the NUM had to persuade the industry to improve the wage rates of the lesser-skilled workers.[41] Because of their numerical preponderance, the Chamber regarded this as an expensive proposition. Had the NUM been a craft union, or representative of the skilled sections of the labour force, wage increases would have been dear but not too taxing on profit levels. As it was, the NUM made wage demands on behalf of a work-force half a million strong, and the wage bill that would result from the increases the union demanded was well in excess of what the Chamber claimed its members could or wanted to spend.

As regards the wage curve, there were essentially two philosophies within the industry. Anglo American and JCI wanted to flatten the wage curve by giving the poorest-paid workers increases greater than those offered to the more senior, skilled workers. In line with this, the lowest-paid workers at Anglo American and JCI mines earned significantly more than the wages paid by the other mining houses.[42] Gold Fields and Gencor, on the other hand, believed in keeping the differential between unskilled and skilled constant by applying the same percentage increases across the board. Again, as a result, skilled workers at Gold Fields and Gencor tended to earn more than those elsewhere. The other mining houses – Rand Mines and Anglovaal – fell somewhere between these two positions. Thus, upon opening wage negotiations in 1986, the NUM faced a Chamber divided over wage philosophy. A year before, Anglo American had broken ranks with the Chamber and made a separate offer to the union, which had been accepted. Division within the Chamber over wages in 1986 was, therefore, nothing new.

At the start of the negotiations, the union demanded a 45 per cent wage increase for 1986, and after the fourth meeting of the parties the Chamber offered 17 per cent for the lowest categories and 12 per cent for the highest. This offer was consistent with the union's attempt to raise the welfare of the worst-paid workers, but it did not consider the increase enough.[43] Consequently, the union rejected the offer; when it declared a dispute, a conciliation board was appointed. Finally, a settlement was reached in October, without a strike being called. Union leaders regarded 1986 as a successful year; wages had improved, so the union believed, because of its actions, and without a serious test of strength.

A momentum for major change had now developed. To announce its intention well in advance, the union sent a letter to the Chamber in March 1987, in which it demanded a 55 per cent increase for unskilled and semi-skilled workers, and a 40 per cent increase for skilled workers. While the demand was, again, consistent with its concern for the least-paid workers in the industry, the irony of the situation was that the union was most strongly represented in the mining houses where the least-skilled workers earned the best wages, and weakly represented in mining houses where the least-skilled workers earned the worst. Indeed, Anglo American's lawyers claimed during the arbitration hearings held later in 1987 that the union used its strength at their client's mines to pressure other houses to change their wage curves.[44]

The parties met on 15 May to negotiate, well ahead of the normal bargaining period. At the conclusion of this meeting, no settlement was reached, and a second meeting took place two days later, at the end of which the union declared a deadlock, and passed a resolution calling for a conciliation board. On 14 June the Minister of Manpower approved the application for a board, which met on 25 June, but lack of a settlement made inevitable a second board meeting, held on 30 June. At this meeting the NUM rejected the Chamber's offer, called for a strike ballot, and by mid-July, after the passage of the requisite time, it declared a legal strike.

In reaction the Chamber accused the union of bargaining in bad faith. It declared that the union had 'made a mockery' of the negotiation process, 'did not negotiate seriously' and was 'merely going through the motions to reach a formal deadlock' so that it could be 'in a position to threaten and eventually to stage a lawful strike'.[45] For its part, the NUM believed that there was a great deal to be gained by a test of the workers' strength. Perhaps it expected Anglo to break ranks and make a separate offer, given that most of Anglo's mines were seriously affected by the strike. At least the Chamber, which had shown such strain in the early 1980s over the wage question, could be expected to divide.

Gavin Relly, chairman of Anglo American, summarised the mood of the Chamber at the time when he noted with disquiet at a shareholders' meeting that the NUM had gone far beyond what could be considered reasonable. He and his colleagues did not agree with the NUM leadership's political views on 'nationalisation, socialism and sanctions', noted that the strike placed a serious burden on the mine companies, and threatened that strikes of this kind would persuade management more and more to substitute capital for labour. It was not that they were against 'independent free trade unions', Relly argued, but against the undesirable direction this trade union was taking.[46] Similarly, Naas Steenkamp of Gencor, who led the Chamber's wage-negotiating team in 1986, claimed that 'the point has now been reached where the future viability of the industry is at stake'. Steenkamp insisted that 'high wage increases across the board' could not be granted 'without any matching increases from the labour productivity side'.[47]

Since members of the Chamber believed that their offer was not only fair but generous, a reconsideration of wages was simply out of the question. While the Chamber reconsidered other issues, and eventually offered revised accident, illness and leave-allowance schemes, it was not prepared at any time to change the wage offer. A week into the strike, Vaal Reefs, an Anglo-administered mine, imposed a lock-out at its number 6 shaft, and thus set the way for an unprecedented use of the lock-out as a strike-breaking device, resulting eventually in mass dismissals on a large scale. Vaal Reefs' defence was that some of its shafts were low-grade, marginal operations, and that they would have to be shut down unless work was resumed soon, within a few days. The NUM's lawyers saw this, however, as an underhand threat, and argued that the use of the lock-out constituted an unfair labour practice, as its purpose was to effect cheap retrench-ment. Other shafts at Anglo mines gave similar notice of a pending lock-out, and when workers did not return within the specified time, mass dismissals began at the end of August.[48] All told, 50 000 African workers were dismissed from their jobs.

In the context of the wider labour market, the strategy proved successful. Retrenched and fired workers were soon replaced by thousands of willing and desperate work-seekers from the homelands and foreign labour-supplying states. Indeed, during the strike employment queues at TEBA recruiting stations significantly in-creased. Lined up behind Anglo American, the industry compelled the union back to the bargaining table. When Anglo American began to dismiss workers, the union faced up to the inevitable and com-municated its willingness to re-open negotiations. This time, in August 1987, the Chamber offered the original wage package, but with revised benefits.[49] The wage offer, declined two months before, was accepted.

The 1987 strike and management's response considerably damaged the union. The NUM claimed in fact that mine management had deliberately and systematically sought to undermine its ac-tivities. Anglo American, on whose mines the union was strongest, was particularly keen on stifling the union, a union report declared.[50] It noted that union recognition was withdrawn from five mines after the strike; that shaft stewards were increasingly marginalised in day-to-day industrial relations affairs; that management no longer allowed the holding of union meetings on mine property; that office facilities (and access to telephones) had become more restricted; and that union members and officials were subject to systematic victimisa-tion and harassment.[51] Partly as a result of the blows it had received, the NUM settled with the Chamber in subsequent wage negotiations between 1988 and 1990, and did not call its members out on strike.

Considered as a whole, wage negotiations between the NUM and the Chamber during the 1980s made in fact only a marginal impact on the real earnings of African workers. In cash terms the increases appear significant, wages having risen by R3 233 between 1982 and 1987, compared to R1 417 and R726 between 1976 and 1981, and 1971 and 1975 respectively (periods before the NUM existed). If 1970 wages are set to 100, the indexed cash wage increased by R1 555 during the years of the union's existence, compared to R682 and R424 for the pre-union periods. Converted to real terms, however, the increases were less dramatic. The purchasing power of real wages increased by a mere 18,6 per cent in the union period, whereas in the pre-union periods they improved by 66 per cent and 289 per cent respectively. The fact that base earnings in the pre-union period were extremely low should be kept in mind when making these comparisons. Nevertheless, the NUM proved unable to drive up real earnings during the period of wage-bargaining. Indeed, between 1984 and 1987 real wages actually fell.

Table 6.2. Gold price and African wages, 1950–1987 (in annual rands).

Year	Gold Price Rands	African cash wages Rands	African cash wages Index	Real wages Index
1950	24,82	104,6	50,3	92,3
1955	25,03	131,8	63,4	91,9
1960	25,07	143,2	68,9	90,3
1965	25,09	176,1	84,7	100,2
1970	25,84	207,8	100,0	100,0
1971	28,64	221,2	106,4	100,6
1972	39,66	256,7	123,5	109,3
1973	65,08	349,7	168,3	135,5
1974	107,42	564,8	271,8	196,2
1975	111,62	947,9	456,1	289,6
1976	103,77	1 102,8	530,7	303,4
1977	125,10	1 235,0	594,2	305,2
1978	168,90	1 420,5	683,5	314,5
1979	254,85	1 668,6	802,9	324,9
1980	479,57	2 037,4	980,3	347,0
1981	402,61	2 520,0	1 212,7	369,3
1982	412,41	2 985,1	1 437,0	384,4
1983	472,95	3 435,8	1 653,4	393,5
1984	525,45	3 975,0	1 912,9	407,6
1985	702,04	4 452,0	2 142,4	392,9
1986	837,28	5 127,0	2 467,3	381,5
1987	904,26	6 218,0	2 992,3	403,4

Source: Hirschson (1988).

Recognising this, the NUM declared that in the area of wages 'progress has been slow and inadequate'. Consequently, in April 1989, the union adopted an explicit wage policy, which reaffirmed a commitment to 'raise wages of all workers, but especially the lower paid'. It called for progress towards the paying of a 'living wage', a national minimum wage, 'good' annual wage increases, reduction of wage gaps, and higher pay for underground work.[52] Moreover, the NUM wanted to standardise wages for all mines, and to achieve a minimum wage of R600 per month for the least-skilled worker, and a single, universal wage curve for the entire industry. These demands made clear that the NUM's achievements in bargaining with the

Chamber during the 1980s had at best been partial and that the Chamber still held the upper hand.

Incorporation into Politics

In support of its more work-related goals the NUM embraced other mechanisms during the decade, including openly political ones. Union leaders did not believe in any separation between workplace and broader politics, and used the latter to offset the lack of progress in the former. As is well known, sanctions and disinvestment against the state were actively promoted by the union. At the NUM congress of 1987, the union gave support to 'all forms of international pressure' against the state, as a means towards the building of a 'democratic and non-racial' society, and as mechanisms by which the state could be undermined and apartheid eroded.

The NUM also embraced the concept of one-person one-vote in a unitary and united South Africa. In 1987 the union formally adopted the Freedom Charter, the guiding historical document of the African National Congress, as the basis for its own policies. As part of the larger labour federation of COSATU, the union established a formal political alliance with the ANC and the South African Communist Party in 1991. With the ANC, the NUM began to advocate the nationalisation of the mining industry to redress inequality between white and black. Ramaphosa believed that the state should play a leading role in redistributing wealth, in the context of a planned, centralised socialist economy.[53]

The union, therefore, came to participate directly in the struggle over state power in South Africa, and identified strongly with a particular set of political constituencies. In doing so, the leadership hoped to shape the political processes involved in negotiations between the ANC and the National Party about the terms of the state's restructuring, and therefore the resulting, but as yet indeterminate, state policies regarding the mining industry.

Compounds as Contested Institutions

A Total Institution

In 1985 the vast majority of African mine employees – almost 98 per cent – were housed in single-sex accommodation known as mine compounds or (more recently) hostels. Compounds are barrack-like structures located close to mine shafts, and they service the labour needs of individual mines. Historically, living conditions within the compounds have been crowded, lacking in privacy, and often primitive. But in the 1970s some of the mining houses began to upgrade their compound facilities. Members of management recognised that the poor conditions of the compounds were part of reason inter-group violence among African workers was increasing.[1] In addition a number of academic studies highlighted the primitive conditions prevailing in the compounds, and caused the mining houses considerable embarrassment.[2] As a result, significant sums of money were invested in upgrading compound facilities and services, and those new compounds built in the late 1970s and 1980s were provided with more modern facilities and improved services.

The upgrading of the compound did not, however, alter its social character. Compounds are instruments which enable management to regiment and mobilise labour. Because of their typical location within walking distance of the mine shaft, African miners can be called to shift *en masse*, at the most awkward of times, around the clock. Whereas, in the case of the industrial factory, management has to rely on workers' dependence on wages and sense of discipline to put in a punctual appearance for the working day, the compound obviates this concern, as the residents can be collectively roused and called to shift.

A related benefit of the compound is the ability it offers the mine industry to feed the work-force *en masse*. Charles van Onselen has pointed out in his study of African labour in the mining industry of Southern Rhodesia how the mine compound granted management a degree of control over the eating habits of its work-force that was not possible and indeed unheard of elsewhere.[3] Mining is hard, physical and exhausting work. By regulating diet and food consumption, management hoped to maintain a physically fit labour force. It was not that it was a cheaper form of food provision, as Francis Wilson pointed out, but it gave management control over the diet and therefore the physical well-being of the miners.[4]

Since its origin a century ago, the compound has insulated and sheltered the mine labour force from politics. In the case of the 1920 and 1946 African miners' strikes and the mine disturbances of the 1970s, mine management insisted on keeping trade-union organisers and political activists out of the compounds. Gate-keeping the compounds in this way helped to separate the miners from a developing political culture of resistance in the townships of South Africa. It was only during the 1980s, however, as this chapter will show, that politics began to penetrate the compound and African miners came to be systematically incorporated into the mainstream political life of the country.

In the past, when the need arose, compounds were easily converted into jails. During the 1920 African miners' strike, so Frederick Johnstone tells the story, 9 000 African miners in four mines refused to work, and 'in response to this, the army was called in, and it surrounded and entered the compounds, and the Director of Native Labour ... gave the striking workers the choice of returning to work or being arrested'.[5] Dunbar Moodie has documented in the case of the 1946 African miners' strike the ease with which the compound was transformed into a jail, a fact which, combined with security pressure, broke that particular strike.[6] In later periods as well, the physical and political architecture of the compound made possible its transformation into an instrument of coercion. This was indeed precisely part of the purpose behind its construction and, later, its modification.

Studies of the social dynamics of mine compounds have on the whole tended to focus on their undoubted managerial advantages. There exists a clarity within the literature about the nature of compounds as creations and instruments of managerial authority. Earlier marxist studies were seduced by this, and constructed an image of the compound as a tool of managerial control, in which workers were seen to be the pliable victims.[7] John Rex went so far as to argue that the compound was one of three South African institutions which made up a perfect system of labour regimentation.[8] Later studies have, however, sketched a more nuanced picture. In these studies worker resistance and the limits of managerial power over the compounds are taken into account, and the compound is seen to belong to a family of total institutions, much like a prison or mental hospital.[9] Though a considerable advance on the earlier work, this perspective nevertheless has supported the notion that compounds serve the interests of management.

The present chapter gives an account of how compounds have come, on the contrary, to serve the interests of organised African workers. During the 1987 miners' strike, groups of African workers used the total character of the compound to enforce worker solidarity, taking control of the compounds out of the hands of management and thereby subverting and thwarting managerial authority. Members of the strike committees kept strangers out of the compounds, and at the same time enforced the strike by keeping workers in. Though worker control of the compounds was firmly established at only a few mine shafts, and was temporary, lasting at most for about four weeks, nevertheless the experience of worker control during the strike showed how total institutions could be used against management to serve worker-based interests. It also made a lasting impact on management styles. At some of the mining houses that had been affected by the strike, management would in the aftermath propose co-operative structures of compound governance, put forward worker-participation schemes in their everyday running, and began consulting with the NUM about spheres of co-operation. Little may be certain about where the proposals for worker participation will lead, but one can confidently say that life in the compounds will never quite be the same again.

A Note on Methodology

Three mines and the compounds serving a number of their shafts form the empirical basis of the studies in this chapter. They rely for their data on the great deal of information that emerged from the legal proceedings regarding the mass dismissals of African workers after the August 1987 strike. The arbitration proceedings,[10] as well as material released as a result of the legal process of discovery before settlement, comprise the data-base of this chapter. While the documentation of the arbitration proceedings falls within the public domain, the material released to the lawyers of both of the parties involved in the dispute does not. The author was privy to that information, but a condition attached to consulting the material was that it could not be directly cited.

The three mines which featured most prominently in the arbitration proceedings were the Vaal Reefs Exploration Company (situated in the Klerksdorp area of western Transvaal), Western Deep Levels (Far West Rand area) and Western Holdings (Orange Free State). All three mines are administered by Anglo American Corporation, and are large employers of labour. In 1986 Vaal Reefs employed 44 912 African workers, Western Deep 24 972, and Western Holdings 39 634; together they constituted 63 per cent of Anglo American's African labour force, and about 25 per cent of the gold industry as a whole.[11] For reasons discussed in the previous chapter, Anglo American mines have tended to be well unionised. Altogether 56 per cent of the labour force at Western Holdings, 48 at Western Deep, and 53 per cent at Vaal Reefs were unionised in 1986.[12] Their compounds were not only large, but also the site of intense union politics.

Because the availability of source material dictated the mines and compounds studied, the conclusions drawn here cannot be generalised with any degree of certainty to other mines and compounds. As already noted, the compounds studied were unusually large and unionised, and were probably at the forefront of struggles in the compounds rather than typical of them. Methodologically, the procedure whereby a scholar bases research on a sample of cases readily at hand rather than on what is compelled by principles of representativeness is known as convenience sampling. Such a procedure is justified when free and open access to source material of those cases selected under a representative sample is denied or made difficult. This study of compounds, therefore, does not pretend to be representative of the industry as a whole, but it is likely that some of the dynamics described here are typical of compounds beyond the cases studied.

Compound Supervision under Strike Conditions

Mine compounds fall under the administrative authority of mine management. At Anglo American mines, a compound manager (usually white) is responsible for individual compounds, and he is assisted by unit supervisors, prefects and gate guards (most of whom are African). The compound manager is in turn accountable to a community-service personnel superintendent, one of three senior mine superintendents. In the past, supervisory structures were organised on an ethnic and 'tribal' basis, but this system was formally abandoned by the early 1980s.[13] NUM leaders claim that the new bureaucratic system nevertheless reproduces the earlier ethnic-centred one.

The NUM enjoyed access to its members in the compounds on the basis of agreements established with mine management in the early 1980s. During the build-up to the 1987 strike, union leaders sought therefore to ensure that management did not disrupt or sever their contact with union members. In a letter of 6 August 1987 to the president of the Chamber of Mines, Cyril Ramaphosa demanded that management honour previous agreements in this regard.[14] He insisted too that union officials be allowed to retain their normal rights of access and asked for assurances that meetings could be held in the compounds. With such a large membership spread over so many mines and compounds, constant, direct communication was seen to be crucial to the success of the strike. In response, the president of the Chamber saw no reason why such special assurance had to be given to the union.

In the first week of the strike, the NUM appointed special strike committees at each mine, made up of union members dwelling in the compounds and representing their residents. The committee's task was to co-ordinate activities between the compounds and the various levels of union leadership during the strike. This committee became the fulcrum around which the strike was organised and enforced. Members of the union's mine-level branch committees also served on strike committees. In some instances, the two committees were virtually indistinguishable.

At Western Deep Levels, an agreement was initially reached in early August between the NUM branch committee and mine management that compounds were to remain under managerial supervision. It was understood that the union 'controlled the strike' and that management 'ran the hostels'. Four days after the agreement, however, the strike committees took over the compound kitchens and canteens. Members of the strike committee refused to let officials into the compounds, and service and repair personnel were only allowed to enter the compound under their supervision. The compound manager was informed that he no longer had authority in the compound, and that the strike committee would run the kitchens.

There is no clear evidence that can explain the arguably subversive behaviour of the strike committee, for the committee plainly dishonoured the prior agreement between the NUM branch committee and mine management. It is possible, though, that the agreement had been made in bad faith, and that the NUM never intended to uphold it. Indeed, at its annual congress held in April 1987, a resolution imploring workers to 'seize control' of the compounds was adopted – 1987 was heralded as the year in which the working class would 'seize control' of their destiny. In the light of this, one could argue that members of the strike committee were acting in terms consistent with congress resolutions.

It is also possible that once established, the strike committee had developed a measure of independence from formal union structures. At Western Deep Levels, the strike committee was to acquire a reputation for militancy, and in the politics of strike organisation, would usually take the initiative. Strike-committee members were on the whole less restrained than the union's central, regional or even branch leaders desired, but because the committees held much of the power during the strike, the union eventually fell in line with their decisions. Taking control of the compounds can thus be seen as an example of this evolving power structure. Another example would be the way the committees persisted in enforcing the strike even after a final agreement had been reached between the NUM and the Chamber towards the end of August.

Whatever the reason, the strike committees held onto the compounds during the course of the strike. Before long, however, they realised they were not capable of fully servicing the compounds, and turned to the compound managers for assistance. At Western Deep Levels, the compound manager agreed to supply the committee with meal tickets and cleaning materials, and promised to assist with garbage removal, spillage in the kitchen, and electrical problems. He also went along with the strike committee's insistence that the electricians could only enter the compound under escort. What emerged was an unspoken agreement about spheres of authority. The strike and branch committees ran the compound, albeit at times with management's assistance, and though the manager complained about his lost and subverted authority, he did very little to regain it during the strike.[15]

For their part, compound managers feared that if they did not supply the strike committee with items needed to keep the compounds in running order, they might face much more expensive breakdowns later. It was in their interests to keep compound equipment operational, even if it meant co-operation with the strike committee. Furthermore, mine-level management acquiesced in the strike committee's control of the compounds because it was recognised as short-lived, and because this control could possibly work in management's favour. An incompetent strike committee could potentially do more to turn compound residents against the union than any of management's counter-strategies.

Leaving the strike committee with the responsibility of feeding hostel residents was a very good example of how management benefited from the committee's shortcomings. For their part, strike committee members considered food provision as central to controlling the hostels. Control over food strengthened overall control of the compounds, and subverted management's capacity to use food as a device to manipulate workers. The strike committee could not, however, feed the work-force with its own resources, and had to rely on some level of co-operation with management. It relied on management to supply the food and supervise its delivery to the compounds. Since the strike committee did not have access to kitchen staff, the retention of their services was essential. Members of the strike committee hoped that if management provided the food and the kitchen staff, all they had to do was supervise food preparation and service.

At first, with the strike young and the strikers enthusiastic, the strike committee kept its hand efficiently on the kitchens. However, as the strike wore on, the capacity of the committee to maintain this level of performance deteriorated. This was a result of the breakdown of equipment. At Western Holdings (Free State) the computerised feeding system in one compound broke down. At Vaal Reefs, the badge-reading machine and the kitchen computer, both part of the computerised feeding system, would not read the workers' identification cards. Some of these problems were recurrent. Under non-strike circumstances the solution was constant, hands-on servicing. However, as a result of the struggle between members of the strike committee and management, servicing and repair procedures became complicated. Some of the problems were slowly attended to. Often, failed equipment was not repaired at all.

As a result, hostel residents were poorly and erratically fed. This tended to undermine the legitimacy of the strike committee's actions. Seen in conjunction with other aspects of the committee's behaviour, the inability to properly feed and service the needs of hostel residents subverted its moral authority in the eyes of ordinary workers. Arguably this contributed to their questioning the wisdom of the strike action itself.

Subverting Managerial Authority

During the strike the union was faced with the decision of either sending workers home or confining them to the compound. At home, the workers would both remain on strike and be out of harm's way. If they remained in the compound, on the other hand, class solidarity could be forged, but this raised a number of logistical problems. At first the NUM encouraged striking workers to return home, and some did. However, in fear that they might lose their jobs if they left the mines, a majority of the workers remained in the compounds.

During the first week of the strike in August 1987, Vaal Reefs and Western Deep Levels experienced a complete stay-away. While Western Holdings managed to keep one shaft in operation, well-unionised mines, such as those administered by Anglo American, were severely affected by the strike. Poorly unionised mines (such as those administered by Gold Fields) were on the other hand almost fully operational.[16] It needs to be said that the stay-aways on these mines, whether partial or complete, were not wholly voluntary. Some of the workers, especially those most vulnerable in the labour market, were afraid of losing their jobs. A significant number of foreign workers, for example, knew that management had the power to fire them and were aware that in their home countries thousands of work-seekers were lined up at TEBA offices waiting for positions to open.[17] On the sending side, government officials from Lesotho and Mozambique were afraid that future employment opportunities for their compatriots would be threatened if their workers participated in the strike, and a number of state-led delegations came to the mines to appeal to workers to refrain from doing so.[18]

As a result of labour market pressure of this kind, the NUM was unable to count on a universal and voluntary solidarity among workers, and thus faced the need to develop strategies to maximise strike participation, even in the face of resistance. One strategy was simply to prevent miners physically from leaving the compound or entering the shaft.[19] This was resorted to often enough, and to great effect, but it was the union's final weapon, used when all else failed. Management self-righteously referred to this behaviour as intimidation, which on one level it clearly was. But more fundamentally it had to do with the social problem of maintaining a collective solidarity in the face of known and unknown defection, in a situation where the relative balance of costs and benefits of participating in the strike was not the same for all workers. For the union's part, it was relatively costly to use coercion: it brought a bad press as well as a loss of possible support among workers. Moreover, the union lacked the physical capacity to police extensively and deal with every act of deviance. Consequently, even though forms of coercion were used often enough, greater reliance was placed on strategies designed to subvert and hold in check management's attempts to get workers back to work.

One strategy was to prevent management from calling the shift over the public address system. During the first week of the strike, when the level of solidarity among workers was relatively high, the calling of the shift did not seem to perturb members of the union and strike committees. Typically, the shifts were called, and there would be no response.[20] Once, however, worker solidarity began to weaken, and some of the workers began to doubt their actions, the divisive tactics of management took on a new meaning. Yesterday's innocuous propaganda became today's poison. Union members, fearing that workers would respond, grew adamant that management had no right to call the shift, and that doing so was provocative under the circumstances.

As a result, the public address system became a focus of struggle between management and union, and access to the facility came to represent a major goal of both parties. Groups of workers began disrupting the work of those operating the public address system. In one episode an operator reported for duty and called the shift. A crowd gathered in front of the broadcast room and would not let him out. In another episode, union supporters rushed up to the broadcast room, and prevented the operators from entering. At yet another compound the operator was allegedly grabbed by members of the union; mine security was sent to his rescue.

If calling the shift was one cause for dispute, another was the broadcasting of corporate information and communication briefs from regional and mine-level management. Just before the strike began, mine-level management warned workers that they 'will lose wages', 'put your jobs at risk', 'if you strike you will not get more wages', 'if you strike, you will actually lose pay and bonus and will also be charged for board and lodging', and that 'the national executive is not going to lose anything as a result of any strike'.[21] In this way management kept workers abreast of corporate views regarding the strike's progress. Briefs like these broadcast over the public address greatly irritated the union members, who regarded them as a source of disinformation. In one extreme case, the public address system was disconnected and destroyed. In numerous cases, it became simply impossible for management to use the public address system at all. For a brief period, then, management yielded to union pressure and stopped calling the shift and playing the briefs. But having subverted managerial control in this way, the union still needed to mobilise the workers directly in support of the strike.

Mass Mobilisation

During the strike African workers resorted to collective action and mass mobilisation, and it became a daily, even hourly, occurrence to see groups of black workers moving from one compound to another.[22] As a form of collective action, the marching crowd served two social purposes. One was to intimidate management by way of a symbolic demonstration of worker power. The other was to promote solidarity among striking workers and to solve the so-called 'free rider' problem.

The crowd intimidated members of management by activating in them some very powerful and deeply rooted racial anxieties of the African 'mob'. Officials tended to see situations of collective action involving Africans as always harbouring the potential for random outbreaks of senseless violence. This perceived threat had been a fear with a long history. It was a major motive for setting up the Native Grievances Enquiry of 1914.[23] The Mine Natives Wages Commission of 1944 had underlined the importance of racial fears in managerial resistance to African unions.[24] More recently, a study of mine violence sponsored by Anglo American and the NUM in 1985 provided extensive documentation of how whites tended to regard Africans as little more than an undifferentiated, potentially hostile group.[25]

In one episode at Western Deep Levels during the 1987 strike, a group of African workers began to move from one compound to another, in an attempt to promote solidarity among striking employees and to intimidate management.[26] With fears of a 'mob' on the march, mine security was alerted, a helicopter was brought in to circle overhead, extra security personnel were commandeered from other compounds, and Casspirs (military vehicles used for riot control purposes by the South African Police, some of which were purchased by the mining houses) were alerted. Little of the repressive technology was used, nor was it in fact at all necessary. The point of the exercise had simply been to intimidate management, which it succeeded in doing. The tactics were right: move around in a large and growing group, sing and wave the flag, and the fear of the African mob brought out all the security hardware management could muster.

Another function of mass mobilisation during the strike was to attain good attendance for union meetings, and to deal with the so-called 'free rider' problem. In the sociological literature, free-riding occurs when individuals do not participate in the collective effort but benefit from its consequences.[27] Since, if everyone believed that others would do the work on their behalf, there would be no collective action, groups will always be inclined to minimise the 'free rider' problem, by incorporating as many participants as possible in the collective action. Mass mobilisation thus serves as a form of moral suasion.[28]

During the 1987 strike, just before meetings were held, groups of workers would go from compound to compound, singing and with flag flying, collecting more and more workers on their way. Under these circumstances, the power of the disapproving crowd could not be resisted, and any free rider would feel obliged to join the frequently held meetings.

During the strike, then, the union managed to organise a great show of support among compound residents, sufficient to intimidate management and win over dissenters among the ranks of the workers. This did not mean, however, that its control was paramount.

Policing the Strike

In the final analysis, the real power of the strike committee resided in the capacity to control directly the behaviour of workers resident in compounds, and ultimately to mete out punishment of one sort or another for the infringement of formal and implicit rules. Indeed, from the very beginning of the strike, union members succeeded in regulating the behaviour of non-union members as well as members.[29]

The policing of the strike occurred in a number of ways. The most important strategy was to regulate entry to and exit from the compounds. This involved keeping outsiders (usually management and its representatives) out, and insiders in. Outsiders had to get the permission of the strike committee to enter the compound, and, if granted, would be escorted into and out of the compound. Throughout the strike, all outsiders, be they compound management staff, mine security, service personnel, even the people who delivered food and other goods to the compound, were screened and escorted through the compound. When outsiders nevertheless got in without permission, the strike committee was greatly upset. In one case, two members of mine security were found roaming in a Western Holdings mine compound. Members of the strike committee held the security men in the union's office, and handed them over to management, threatening to lodge an official complaint.

The control over the movement of compound residents, which was the other side of gate-keeping the compounds, was not anywhere as straightforward as keeping outsiders out. The strike committee issued yellow tickets to *bona fide* compound residents, which represented their pass in and out of the compounds. Those who had no ticket but desired entry had to explain themselves, and often were subjected to a body search. Whether or not this system actually worked day in and day out is not clear from the evidence. But by itself, the yellow ticket system was not enough of a device to police movement, and had to be supplemented with additional measures.

The problem was one of scale. The strike committees did not have the resources to monitor all the exits and entrances. To police all of these required a personnel the strike committee and union either could not or would not supply. Indeed, it was never the NUM's intention or business to police compounds, and while some of its members easily acquired the art, they were not schooled nor had the will to play policeman with their comrades.

One way of coping with the scale of policing was simply to reduce the number of possible entrances and exits.[30] At Western Deep Levels, for example, the strike committees tried to reduce the number of entrances by locking the gates. For the committees, keeping the gates thus locked proved a critical part of their regime. In one episode, a member of the strike committee found a gate unlocked, and asked mine security to lock it. When the compound manager claimed that no one was available to lock the gate, the committee member went to the compound, fetched a device and locked the gate. Of course, the locking of the gates did not go unchallenged by management, which considered it contrary to mine policy and practice. For this reason many a gate was unlocked by mine security. This underscored the fact that the union's control over gate-keeping was incomplete, and that management at times regained control. On one occasion, for example, a group of workers from one compound petitioned the compound manager for permission to visit another compound, and this was refused. The workers abided by the decision, thus indicating that the manager still had some say over who went into which compound. Sometimes, too, management aggressively challenged the union's right to guard the gates. At no stage was the union's control over the gates ever complete, and many times members of the strike committee simply lost record of who passed where.

The strike committee also tried to police access to the change-houses and the waiting-places where workers assembled to go underground. But the main area of control was the compound, and on rare occasions recourse to physical coercion was used to discipline the work-force. Unwelcome guests in the compounds were readily thrown out. There were also rumours of physical assault, as in one reported incident when seven shaft stewards allegedly assaulted a person in front of the gate, below the administration offices. How widespread these activities were is not clear from the available evidence.[31] In many cases of assault, individual workers took it upon themselves to mete out punishment. Strike committees, which strictly speaking represented compound residents, often acted on their own initiatives, independent of the union. But it would be false to suggest that the NUM promoted unrestrained violence. On one occasion at a Western Deep Levels compound, were it not for the union's intervention, the compound manager could have lost his life. When it came to other forms of violence, the union also sought justice rather than partisan advantage and co-operated with mine security and the South African Police. In one case, three men were arrested by members of the strike committee at one compound for raping an African woman.[32] The committee handed the men over to mine security, who passed them on to the South African Police. In another instance a man was caught with dagga (marijuana), and appealed to the union for assistance. They were unsympathetic: 'He had asked for help from the NUM but had been told he knew he was in the wrong, and shouldn't have tried to bring it into the compound.'

Yet co-operation of this kind with mine security and the police was occasional only and the general tone of the relationship between the parties tended to be markedly hostile, indicating the fierce contestation for control of the compounds. Though the striking workers seemed on the face of it to have undisputed authority and possession, yet management still maintained a residual toe-hold during the strike which in the aftermath was extended to a substantial presence, if not a replica of its former power.

A Contested Institution

Towards the end of August 1987, a great deal of confusion reigned in the compounds. An increasing number of workers wanted to return to work, and in this were encouraged by management. But the strike committees did not take kindly to what their members thought was a premature ending of the strike. Anglo American, the mining house which took the lead in resolving the strike, first used a lock-out in order to force the union back to the bargaining table, then dismissed workers *en masse* at some of its mines. This move proved successful, and in September the union accepted the same wage offer refused at the beginning of July, despite protest by its lawyers that the mass dismissals were an unfair labour practice.

Before the mass dismissals came into effect the residents at Western Deep Levels, and presumably at other mines affected by the strike, were divided into essentially two groups. One wanted an early return to work. Having received warning of possible mass dismissals, these workers feared that they might lose their jobs, a fate which indeed befell many of them later. As soon as the opportunity presented itself, these workers went underground. On 26 August 1987, the compound manager reported that 'large numbers of black workers consisting of dayshift, nightshift and afternoon shift, entered the crush, reporting for duty'.[33] In other cases, groups of returning workers were given security protection by management. Returning workers were escorted by mine security officials and their Casspir to the mine shaft.

The other group of workers, the strike stalwarts or *bittereinders*, saw the early return to work as a betrayal of the struggle. This group tried to frustrate as best possible management's efforts at persuading others to return to work, by preventing workers from leaving the compound, and keeping the gates locked and the change-houses shut. But the three-week strike was really over. Once the mass dismissals came into effect, the union lost the battle. Though union officials tried to monitor the pay-outs after the dismissals, even here corporate management refused to consider the union's proposals for appropriate procedures.

At the compounds, too, the union quickly lost ground to management.[34] At Vaal Reefs, one manager took back control of a compound, escorted by mine security and a Casspir. Elsewhere, mine security and six 'rescue vehicles' took back a third compound.[35] Compound residents were told over the public address system that security had taken over, and that people were to stay in their rooms and wait for the shift to be called.

As for the workers dismissed, they were packed off home. Those dismissed but remaining in the compounds were told that they had to leave. Generally, dismissed workers were not given leave or re-engagement certificates. Union and strike leaders in particular became targets of management's offensive.[36]

The compounds would never quite be the same again. In the aftermath of the strike, some mining-house officials began speaking about developing more co-operative systems of compound management. At some of the newer compounds, such as JCI's H. J. Joel mine, compound management provided for worker participation in decision-making structures. The struggle over the mine compound during the strike reinforced the resolve of those officials committed to developing alternatives to migrant labour and mine compounds. However, since migrant labour was still expected to remain the predominant mode of organising the industry's labour supplies in the future, there were no plans to abandon compounds altogether. Notwithstanding the changes emerging from the 1987 strike experience, the mine compound is destined to remain an essential part of the labour institutions of the mines.

8
The Struggle over the Colour Bar

Introduction[1]

For much of the twentieth century the advancement of African workers in the mining industry was limited by a pervasive and institutionalised system of job discrimination, known as the colour or job bar.[2] One aspect of this bar was to disallow African apprentices in the skilled trades of the mines, which was consistent with the general pattern in other trades. Another aspect was Job Reservation Determination no. 27 of 1971, which reserved the majority of jobs in sampling, surveying and ventilation in underground work for whites only. The most notable aspect of the bar was contained in the Mines and Works Amendment Act of 1956, a consolidated statute based on the original law of 1911 and on racial amendments made in 1926. This legislation reserved certain key jobs in the mines for holders of 'certificates of competency', and only those persons who were 'Europeans, Cape Coloureds, Cape Malays, Mauritius Creoles and St Helena persons' were reckoned eligible. Of the eleven such certificates the blasting certificate was the most important, as it was the ticket to supervising rock-breaking, the primary operation in gold-mining, and put miners in line for promotion to more senior jobs. Finally, the closed shop concluded with the white unions in 1937, the related Allocation of Occupations Agreement concluded later, and the Better Utilisation of Labour Agreement of 1973 all restricted African access to skilled and senior jobs, though compared to the main aspects of the bar their impact was marginal.[3]

Of the 20 000 white workers hired by the mines in the 1970s, the 10 000 artisans belonging to one of the six craft unions* were well

* Amalgamated Engineering Union, Amalgamated Society of Woodworkers, Amal-

qualified, possessed transferable skills and, if they had to, could find jobs outside of mining. Their competitiveness in the labour market was further strengthened by a closed shop and regulated apprenticeship system. Although prior to its abolition artisans were nervous about the prospect of the disappearance of the bar, they accommodated themselves to its passing. The 10 000 other workers belonging to the Mine Workers' Union (MWU) were, however, relatively insecure in the labour market, and directly threatened by African advancement. Most of these workers had little formal training and possessed skills that could only be used in mining; they relied on the certification system provided by the Mines and Works Amendment Act for their jobs. For these reasons, they actively resisted managerial challenges to the bar. The bar also protected the jobs of a number of white employees who were classified as officials (by Determination no. 27) and represented by the Underground Officials' Association (UOA), one of the mines' three associations of officials.**

The privileged access of white employees to the better jobs in the mines was the visible and intended consequence of the bar. A less obvious effect was the manner in which it tended to bolster the cheap labour system. Mine-owners passed the costs of maintaining the bar – consisting mainly of inflated wages for some categories of white workers – to African workers, by searching for the cheapest unskilled and semi-skilled workers available; and because of the nature of the labour market these were migrants. For political reasons, the Chamber was unwilling to confront the employment security of white workers directly, and therefore targeted all austerity measures on the mines at African workers. Since workers were barred from skilled jobs, management had little incentive to invest in the improvement of their skills, and could live with the labour turnover associated with a system of oscillating labour migration. As a result, the colour bar served to retain African workers in the migrant market and cheapen their labour.

In a number of reports published between 1979 and 1981, the Weihahn Commission recommended that the colour bar in all its various guises be abolished, in line with the state's more general commitment to end racial discrimination in employment opportunities. As a result, in 1983 the bar on African apprentices was lifted and Job Reservation Determination no. 27 withdrawn. The resistance

gamated Union of Building Trade Workers, Iron Moulders' Society, Boilermakers, Iron and Steel Workers', Shipbuilders' and Welders' Society, and Electrical Workers' Association.
**The other two were the Mine Surface Officials' Association (MSOA) and the South African Technical Officials' Association (SATOA).

of the MWU delayed, but did not stop, changes to the Mines and Works Amendment Act, and in 1987 parliament removed its racial clauses. A year later the Minister of Economic Affairs and Technology published in the *Government Gazette* the new colour-blind regulations following from the new legislation, as well as those that specifically affected mine safety.[4] But the appearance of the regulations did not signal the end of a growing controversy between capital and state over the manner in which the legal underpinnings of the bar were to be phased out. For its part, the Chamber wanted as free a hand as possible in shaping the conditions under which African workers were to advance in the industry. In response to an application brought by the Chamber, a Supreme Court judgment struck down the Minister's new regulations in September 1989 on the grounds that they went beyond his powers.[5]

In *Capitalism and Apartheid*, Merle Lipton argued that the abolition of the bar was the Chamber's response to growing skill shortages in the racially segmented labour market.[6] Lipton noted that for this and other cost-related reasons, the Chamber had always been critical of the bar, but that it had been politically too weak to challenge a state which for much of the century was overly sensitive to the interests of white workers.[7] This chapter contests her interpretation. Although the Chamber mobilised arguments in the 1960s about a labour shortage to legitimate limited African advancement, the shortages played an insignificant role in their motivations when the bar was finally abolished in the 1980s. Far more important for the Chamber was a growing managerial commitment to organise work tasks according to substantively rational criteria, at a time when the mines were involved in the reorganisation of job and pay structures. Furthermore, it was not that the Chamber lacked the class capacity to change the bar, or that it was politically too weak to overrule state officials committed to racial privilege. Rather, the Chamber was of itself unable to set in motion a process by which it could both legitimately remove the bar and cope with the class conflict generated by white worker reaction to African advancement. In the 1980s, reformist state officials, who were distant from the interests of white workers, provided the political room and institutional opportunity the Chamber needed to end the bar.

Change to the legislative framework governing the bar did not, however, bring about rapid African advancement in the occupational structure. After 1983 African workers became indentured as apprentices, and could work as recognised artisans. A considerable and growing number of African workers acquired blasting (and other)

certificates. Africans were also appointed to jobs as officials in surveying, sampling and ventilation work. Overall, however, they made up
only a minority of those employed in skilled and senior positions in
the mines. In fact, a number of considerations made it extremely
unlikely that the end of the bar would result in rapid African advancement. Whites were not vacating positions in the occupational structure at a rate at which Africans could fill them rapidly in large
numbers. Moreover, one of the notable features of the post colour-bar
legislation was that it still contained mechanisms to regulate the entry
of Africans into skilled work. So while by 1990 the colour bar had
given way under various pressures, the advancement of African
workers which this allowed proved much slower and more
problematic than had been expected.

The Erosion of the Bar

Historically speaking, the operation of the various aspects of the
bar was never rigid or inflexible. Under the right circumstances and
with appropriate rewards, those white employees protected by the
bar proved, particularly after 1960, willing to bargain over the racial
allocation of work tasks. In this process they ceded more and more
responsibilities to African workers, and unintentionally allowed a
small but significant measure of African advancement, and thus the
erosion of the bar, to occur.

The first attempt at shifting the boundaries of the bar took place in
the 1960s, when the mining houses participated in the so-called
'mining experiments' of 1964–5.[8] At the twelve mines involved in the
experiments, senior African workers were upgraded to the status of
a 'scheduled person', to perform tasks previously done by whites. The
industry was stagnating, as a result of a low and fixed price of gold,
as well as rapidly rising working costs. The experiments were to
ascertain whether improvements in productivity could be attained
with a few modifications in the labour process, including ones that
affected the operation of the bar.

In exchange for improved wages and conditions of service, white
workers ceded some blasting responsibilities to African 'boss boys',
and were left with nominal inspection duties. This only served to
confirm what was already happening in practice, however randomly.
Before the experiments were introduced, Africans had already become involved in blasting operations. The Government Mining Engineer (GME) reported 102, 134 and 75 prosecutions in 1962, 1963 and
1964 respectively for violations of the bar clauses of the Mines and

Works Amendment Act, because of Africans performing work they legally were not entitled to. For the duration of the experiment, however, the GME relaxed the bar clauses of the legislation.

Leaders of the MWU initially went along with the experiments in 1964–5 and the relaxation of the bar.[9] Though its president, Eddie Grundling, believed that members stood to gain sufficiently in wages and benefits from the proposed adjustment in tasks, many of the workers felt directly threatened by the experiments. In the process, as was feared, the number of white miners required underground was halved. Despite the fact that the Chamber had promised the MWU that workers would not be retrenched but employed elsewhere, many workers were dissatisfied with the Chamber's assurances, and a rebel group came out strongly and militantly against the experiments. The group campaigned against Grundling, called out workers in wildcat strikes, and ransacked local union offices. In a major leadership shake-up, Grundling was replaced by Arrie Paulus, who later rose to prominence as a staunch white-supremacist labour leader.

For their part, mine-owners reacted favourably to the results of the experiments. Productivity improved, especially when the experiments were combined with the new blasting and stoping techniques – 'sequential firing' and 'concentrated stoping' as they were known in industry parlance, both introduced in the early 1960s.[10] The changes in the organisation of work and employment underground brought also considerable savings. However, in the light of the negative response of white workers in the MWU, parliament felt obliged to appoint a commission of inquiry into the events. Chaired by F. J. Viljoen, the commission concluded that although the experiments were successful from the point of view of the Chamber and mine management, there was reason to fear the 'wider implications' of relaxing the bar. On the recommendation of the commission, the experiments were therefore terminated.

The experiments had sought to improve the performance of the gold industry at a time not only of relative stagnation, but also of major losses of white labour to other sectors of the economy. As the Wiehahn Commission later documented, the industry lost 23 per cent of its white labour force in the 1960s. The number of artisans declined by 15 per cent, rock-breakers by 25, general miners by 5, and semi-skilled workers by 46.[11] Many left the mines because of more generous conditions of service in other mining, manufacturing and service industries, at a time when these were rapidly expanding.[12]

Although the white labour market was to turn around significantly in the 1970s and 1980s, the decline of the 1960s was regarded within

Chamber circles as the beginning of a serious labour shortage. In 1970 the Chamber's president concluded that 'the mining industry experienced a persistent shortage of white labour ... and there is no reason to expect a significant alleviation'.[13] Between 1968 and 1972 it was recognised that the shortage of white labour posed a serious problem, that it appeared to be getting worse, and that something had to be done about it.[14] Failing this, the shortage would reach a point 'where it places a brake on new mining developments'.[15]

The Chamber's response to the apparent shortage was not to stimulate the white labour market, but to use it to legitimate a measure of African advancement. Relatively feeble efforts were made to draw more whites into mining. The Chamber launched public relations campaigns aimed at white secondary schools, where the attractions of mine employment were promoted. Greater grants-in-aid were awarded to university and college engineering, geophysics and chemistry departments. The recruiting of skilled workers from overseas was also stepped up.[16] But the Chamber was unwilling to offer any wage or other material incentives, for that would have made expensive labour even more expensive.

On the grounds that there were simply no whites to be had for the available jobs, the Chamber put African workers into jobs where there was a declared shortage. Training of African rock-breakers began in the late 1960s: under exemptions granted by the GME, African miners were permitted to handle rock-breaking tasks, albeit under the supervision of white miners. The level of responsibility granted to African workers was, however, not as substantial as in the case of the 'mining experiments', when they had been allowed to handle explosives and drive underground trains.[17] From 1967 onwards, the GME regularly granted exemptions so that Africans could be used in place of whites for blasting and rock-breaking.[18]

The mines also began to train Africans as artisan aides in the early 1970s. Africans could not be employed as artisans on the mines because of the discriminatory practices of apprenticeship committees in the trades as well as the operation of the closed shop. By the early 1970s, however, new technical schedules were drawn up for artisan work, and by 1973 the existing unions agreed to the introduction of a new category of 'artisan aide', and at the same time to the enrichment of their jobs under a Better Utilisation of Labour Agreement (BULA). Drawn from the ranks of existing African employees, this category soon grew to 11 000 workers, becoming the third best-paid job for African workers on the mines, and a popular promotion route for the advancement-blocked African labour force.[19] The artisan aides worked under the direct supervision of white artisans, and could not do any work other than that negotiated between the Chamber and the white union concerned. White workers were promised at the same time that 'there is no intention that European employees will be replaced by the Bantu', and were reassured that 'no European employee will be retrenched as a result of ... the changes proposed'.[20] In exchange for the recognition of the new job category, all white workers who belonged to any of the unions and official associations were to receive a R50 increase on their standard rates of pay, a R50 'responsibility allowance', a R25 increase on leave allowance, and more generous leave arrangements, medical and pension schemes.[21]

At one point, the colour bar was even strengthened. In 1965, the Chamber discussed with the UOA the possible utilisation of African workers in job categories where whites were in short supply, mainly in surveying, sampling and ventilation work. The Chamber reassured the UOA that no changes would occur here without prior discussion. However, in 1969 a small number of junior members of the UAO complained about African 'infiltration' into sampling, surveying and ventilation. On their behalf, the MWU lodged a complaint with the Minister of Labour, who called an Industrial Tribunal. This recommended after its investigation that all but the most junior positions in the departments of sampling, surveying and ventilation be set aside for whites under Job Reservation Determination no. 27.[22] Although the jobs in question were represented by the UOA, the bar determination came as a result of the activism of the MWU, a fact which later made its repeal easier to achieve.[23]

In 1977 mining regulations were changed once again to allow African miners greater scope to perform 'scheduled work', as part of a bargain with white workers struggling to obtain a five-day working week. Both the artisan and production unions had been urging the Chamber to change the six-day working week to five days in the 1970s, a demand first made in the late 1940s. When the Chamber demanded more job fragmentation in return, however, the artisan unions retreated from their demand.[24] The MWU, much more willing to push matters to a strike, accepted the compromise solution of an eleven-shift fortnight, in exchange for modifications in mining regulations regarding 'scheduled work'. As a result African miners now acquired greater responsibility in the labour process, as they could enter faces that had been blasted and charge up drill holes with explosives. White miners still maintained some semblance of ultimate supervision over the African work-gangs.[25]

As a result of all these negotiated changes in the racial allocation of work, the bar became seriously eroded. African workers, though formally precluded from any skilled work, had, as rock-breakers and artisan aides, begun the climb into the ranks of the skilled. When the Wiehahn Commission sat to consider the bar in the late 1970s, these initial steps brought it to the conclusion that the bar was indefensible and that it would be irrational to insist on its retention.[26]

Rationalisation of Work

Of overriding importance to mine management in the 1970s was the absurdity of allocating work-tasks according to racial status rather than standards of efficiency. Indeed, that the managerial mind focused attention on the bar in the 1970s was no accident. African wage levels were increasing rapidly and the range of jobs they had come to occupy was expanding. In return, mine management sought higher levels of labour productivity and a rationalisation of the occupational structure.

Among other problems created by the bar, two in particular stood out. The first had to do with the job content of those white workers who had over the years ceded their tasks to African workers.[27] Many of these workers had assumed more supervisory roles over time, and moved from blue-collar to white-collar work.[28] But a portion of the remaining work-force became visibly underemployed, to the point of possible redundancy. A number of academic studies have observed that a striking feature of the labour process in the 1980s was the nominal roles of some white workers. Dunbar Moodie described how some white workers underground came to 'withdraw' from any labouring activity.[29] Michael O'Donovan wrote that the white worker had come to play merely a sentry role, watching who entered and left the stope, 'a role that may be defunct'.[30] As corporate studies revealed, management was becoming concerned about the degree of underemployment and possible redundancy which the bar had created.[31]

The second concern was management's inability to replace the dearer white workers with cheaper African ones even though the latter were capable of performing the same work. As indicated in Chapter 5, the stratification of the African labour force shifted in the 1970s from an essentially unskilled and manual pattern. Increased experience and ongoing bargaining between the Chamber and the established unions produced significantly more semi-skilled and skilled African workers. As a result, there were many African workers capable of performing work denied them under the colour bar.

In 1975, when some Anglo American mines introduced 'drilling jumbos' and 'raise and blind-hole borers' – machines used for shaft-sinking and development work – management found itself hamstrung by the bar. A report on mine mechanisation at the time suggested that 'we should maintain the principle that only artisan operators should be employed. In this way it will be less troublesome to arrive at our ultimate aim of employing blacks as operators supervised by a white artisan operator.'[32] The report observed that 'blacks are doing nearly all the operating work. Very little operating work is being done by whites.'

Underemployed but protected white workers, and upwardly mobile but blocked African workers, together made up an anomalous occupational structure produced by decades of the operation of the colour bar. Anglo American had in fact already recognised the problems the bar created in the 1960s, when the mining experiments were introduced. At the time, the most pressing problem was that existing job definitions did not cater for the range of jobs African workers had come to perform. Consequently, Anglo management formally evaluated their work and, based on the results, introduced a three-band system, one each for supervisory, clerical, and manual work.[33] It was the first time that 'African work' received such close definition.

All the same, this system was devised firmly within the parameters of the colour bar,[34] and failed to make any linkages between the work performed by Africans and that performed by whites, regarding these as virtually autonomous regions of labouring activity. In the early 1970s, when the range of tasks African workers performed had expanded once more, the three-band system became an inadequate framework in terms of which new jobs could be described and ranked. It was thus an opportune moment to introduce a new job evaluation system consistent with the desire to discard racial discrimination in employment practices. What Anglo American management wanted to do was introduce an integrated occupational structure blind to race, organised on the basis of production and efficiency considerations.

Between 1971 and 1973, therefore, Anglo American introduced a more complex job evaluation scheme, in line with policy. In the process the Corporation could call on prior experience of a racially blind system. In post-independence Zambia, for example, where Anglo American owned a number of copper mines, job structures had evolved along non-racial lines; while at one of its mines in Swaziland, Anglo American had experimented with a more complex job evaluation scheme – the so-called twenty-six point system – and was greatly impressed by its flexibility. At the same time, the industry as a whole began to investigate the possibility of developing a comprehensive and widely applicable job evaluation and grading system. For one thing, the Chamber was interested in developing one for all of its member mines. Moreover, there was growing managerial interest within many mining houses in a device that could be used to assess 'white' work and rationalise job structures along non-racial lines. When the Chamber rejected Anglo American's twenty-six point system as inadequate for the task, officials at Anglo and elsewhere started looking for another job evaluation scheme.

Of a list of four, Anglo American chose the so-called Paterson system. It was elegantly simple, based on a single criterion – that of decision-making authority – for assessing and grading jobs; it correlated strongly with some of the existing job categories; it had an international credibility, and was methodologically robust. By 1975, Anglo American had Paterson introduced for all its African workers in all its mines, and the rest of the industry followed suit. Towards the end of 1978 an attempt was made to extend Paterson to white workers,[35] and by the early 1980s a comprehensive non-racial job-grading scheme had been developed for the whole industry.

Not surprisingly, there emerged a number of serious problems with extending the system to white workers. The MWU objected on the grounds that it meant deracialisation of work by stealth.[36] The MSOA claimed that it was another means by which the Chamber would sneak in a cheap labour system by the back door, in so far as African workers would be substituted for whites in new jobs at lower rates of pay.[37] As Perold pointed out, Paterson was management-based, and did not depend on the co-operation of workers or their unions in order to be implemented. All these objections to the system worked to delay its implementation. Only in 1984 did the MSOA come to an agreement with the Chamber over job evaluation, and even then the agreement was short-lived.[38]

In the way of the rationalisation of work stood the colour bar. Paterson could only in fact be fully implemented with the abolition of racial strictures on employment practices and fuller managerial control over the allocation of work tasks. Thus in the 1970s, when rising African wages began to put pressure on overall working costs, management began to push even more strongly for the removal of the bar, as part of its overall rationalisation of work.

State and Capital

In 1977 parliament appointed two commissions of inquiry – the Riekert and Wiehahn Commissions – that were to transform the relationship between the state and African labour markets in the 1980s. Shortly after their appointment, the Chamber urged the Wiehahn Commission to include the mining industry in the scope of its investigation. Wiehahn had apparently assumed that the special legislation governing the mining industry excluded it from the scope of its research.

In the eyes of the Chamber, the commission provided a long-awaited and propitious opportunity to pursue the interests of mine-owners within a framework legitimated by the state. For the Chamber did not want to confront white workers and their employment security directly. It preferred instead to see the class conflict over the colour bar mediated by the state, and its demands legitimated by the authority of a supposedly neutral, quasi-state body. The removal of the bar by act of state rather than class was certainly regarded by the Chamber as preferable.

But the Chamber saw the commission as an opportunity to challenge not only the colour bar – although that was key – but a range of other related issues as well. As we have seen in Chapter 6, the commission became a vehicle by means of which the provision of trade unions for African workers was adopted as state policy. Furthermore, the Chamber pressed for an industrial council, and an end to closed shops, and encouraged the emergence of a deracialised job structure. Although most of these goals were not attained in the period covered by this study, the quest to end the colour bar should be seen in the light of the broader concerns of the Chamber and the mines it represented.

After a decade of complaints about a white labour shortage, the Chamber in its submissions to the Wiehahn Commission in 1977 conceded that its arguments against the bar had little to do with shortages. 'It is quite evident', noted the Chamber's drafting committee, 'that the real influence of legalised job reservation has been relatively small when viewed statistically.'[39] The committee's arguments rested on future labour needs. 'If one looks into the future the industry will need more skilled labour, which whites and immigrants cannot supply.' On this basis 'the projected demand for skilled labour ... demands that more non-white workers be absorbed into the labour pool'. Thus, 'the mining industry contends that legislated job reservation based on racial discrimination is no longer defensible' because it had limited and would limit the 'optimum use of labour'.

In practice it was not as if there was either a serious need or a desire to employ Africans in areas restricted by the operation of the bar, particularly those involving the blasting certificate. Senior executives of Anglo American, a mining house that played a considerable role in the formulation of the Chamber's evidence, believed that the Corporation was 'not prepared in any way for the issue of blasting certificates to blacks – organisationally, technically, socially or even the training or housing aspects'.[40] They agreed with Gencor's view that it made no sense 'to alienate the MWU for the sake of securing blasting certificates for blacks'. In the Chamber's evidence to the Wiehahn Commission, therefore, the political approach had to be right: the issue of blasting certificates was not to be confronted directly, for that would 'alienate' the MWU, but rather 'what should be done is to get the Commission to consider the definition of the scheduled person'.[41]

Although the commission heard a range of opinions on the colour bar from a variety of quarters, its conclusions were essentially based on the evidence of the Chamber. In the preparation and submission of its evidence the Chamber put considerable effort. Eight days after the commission was appointed, the same committee of the Chamber responsible for submitting evidence to the Franszen Commission, which had investigated and later rejected the five-day working-week in the mining industry, was asked to prepare a draft of views on the bar, industrial relations and an industrial council.[42] Senior executives at Anglo American went to much trouble to ensure that the Chamber's submission was 'really first-class and far-reaching' and that a sound statistical and empirical analysis of employment patterns could be generated for the commission's use.[43]

Because its sentiments were known before the time, it came as no surprise when Wiehahn recommended in its sixth report of 1981 that the state withdraw the racial definition of the 'scheduled person', Reservation no. 27 and the bar on apprentices. To arrive at this, the Chamber and the trade unions were urged by the commission to negotiate the terms for the abolition of the bar subject to the following qualifications: the maintenance of work standards, equal training and experience for appointment to previously 'white' jobs, equal pay for equal work, consultation with trade unions before any changes in existing work practices were brought about, and guarantees of job security for present employees. The sixth report also recommended the formation of an industrial council – a body which could enforce the agreements made between employers and employees – and the replacement of the closed shop and Allocation of Occupations Agreement with a more satisfactory arrangement.[44] These recommendations were accepted by the state in 1981 in a White Paper which approved of the abolition of the bar, subject to the provision that the Chamber and the trade unions reach an agreement about the employment security of white miners 'within a reasonable time'.[45]

The Response of the White Unions

The MWU reacted to the first of the Wiehahn reports by bringing out its members in a sympathy strike in March 1979. White workers at the O'Okiep copper mine in Namaqualand went on strike in protest against the employment of coloured artisans (in what previously had been 'white' work) in line with the new Wiehahn guidelines. But hardline action by the Chamber, combined with the unprecedented refusal of the National Party government to come to the white union's aid, broke the MWU's sympathy strike. The outcome of the strike indicated in no small measure how relations between white mine labour and the state had changed.

Before the 1970s, white miners had formed a significant segment of the National Party's constituencies in mining towns. In the 1950s and 1960s, successive National Party governments cultivated the loyalty of white workers by underwriting white privilege, as the outcome of the 'mining experiments' of 1964–5 partly indicated. By the 1970s, however, the state was coming under considerable pressure from employers to abandon racially protected labour and employment markets. Organised African workers also began to press for reforms in the labour framework during and after the Durban strikes of 1973. Moreover, the 1976 Soweto riots focused the attention of the international community on apartheid policies, including discriminatory and repressive labour practices. Subject to these pressures, state officials began a process of reform, manifest in part by the appointment of the Wiehahn and Riekert Commissions in 1977. Legislation and regulations that discriminated on racial grounds were increasingly withdrawn. As a result, white workers felt abandoned by the National Party, which was seen to be shifting its representation base in favour of the white middle and upper classes. By the 1980s, white workers in mining towns had broken their allegiances with the National Party by joining the Conservative Party, and turned mining towns into white-supremacist constituencies.[46]

In this political context, the Chamber approached the white unions in the early 1980s with a package proposal, in which the issues of an industrial council, the closed shop, and security of employment guarantees were all linked. As indicated earlier, the Chamber wanted agreement on a range of items, not just the bar, for only when such a package had been agreed upon would the Chamber request the state to put through the necessary changes to the Mines and Works Amendment Act. Some of the trade unions were reluctant to participate for one or other reason in the negotiations. Neither SATOA nor UOA was convinced of the value of an industrial council. Both AEU and SATOA pressed for a break-up of the package proposal, the AEU arguing that an industrial council should be introduced first, before any of the other issues were dealt with, while SATOA asked for the reverse.[47] All these disagreements seemed minor, however, compared to the position taken by the MWU, which insisted on the prior rights of white workers. The MWU's stance remained clear from the time of Arrie Paulus's first attack on the sixth part of the Wiehahn Report. There was, he claimed, no shortage of white miners. Paulus believed that the real aim was labour substitution, and that if the bar was taken away, 'sixty per cent of white miners' would potentially lose their jobs and be replaced by African miners.[48]

Government officials hoped that in line with the recommendations of Wiehahn, the parties involved would come to some form of agreement. When the prospect of hoped-for agreement seemed slight, the Minister of Mineral and Energy Affairs, Danie Steyn, publicly announced that end 1985 would serve as a final deadline.[49] When this passed without agreement, the Department of Mineral and Energy Affairs released a typed copy of a draft bill to amend the Mines and Works Act. As an industrial relations consultant of the Chamber later observed, this draft bill acted as a catalyst, prodding the Chamber and the trade unions to sort out an agreement before government unilaterally changed the legislation.[50]

The first draft produced by government was almost to the liking of the MWU, and as a result unacceptable to the Chamber and the majority of the other mine trade unions. Its main feature was the mandatory introduction of a selection board, which could conceivably be dominated by individuals who had a vested interest in limiting African access to certificates of competency. The functions of the board were to lay down entrance requirements and conduct a means test.[51] In terms of this draft bill, the conservative unions could ensure that stringent conditions were laid down for candidates, serving in this way as a pre-training selection process.

Negotiations between the Chamber and twelve trade unions were resumed in January 1987 and, according to Chamber representatives, by March a 'good' draft agreement had been reached.[52] However, both the MWU and SATOA refused to go along with the draft agreement, because they had no faith in the security of employment provisions for the white workers. In spite of this, nine trade unions and the Chamber came to an agreement in July 1986, consisting of three parts. The first part was the formal request to replace the definition of 'scheduled person' with that of a non-racial definition of a 'competent person'.[53] The second part consisted of a draft constitution for an industrial council in the mining industry. The third and final part was a security of employment agreement, which included union demands such as equal pay for equal work, merit-based employment, consultation regarding changes in work practices, and criteria for training and promotion.[54]

In the course of 1986 three further drafts of the legislation appeared. The final draft diluted considerably those provisions the Chamber and signatory unions had objected to. Instead of a mandatory selection board, the fourth draft gave the relevant Minister powers to issue regulations governing the suitability of candidates for certificates of competency, to appoint advisory committees, as well as consult with organisations that represented existing holders of certificates of competency.[55] By this stage, the draft bill evoked only minor criticisms from the Chamber, the artisan unions and officials' associations, which felt that the new, largely discretionary powers vested in the Minister could potentially be applied in a discriminatory fashion.

The Response of the NUM

The leadership of the NUM asked to be allowed to participate in the negotiations over the bar in 1984, but the Chamber refused.[56] The fact was that MWU leaders were not willing to participate in joint discussions with the NUM. Since the Chamber believed that it was with the MWU that the greatest difficulty lay in removing the bar, leaving the NUM out seemed at the time a price worth paying. In fact representatives of the Chamber and NUM leaders had already met to discuss objections to the third version of the proposed bill in 1987, but in the absence of the white unions.

It was quite clear why the NUM wanted to participate actively in the political processes involved in the repeal of the bar. The union represented the interests of a work-force which had laboured in the past under the restrictions of the bar, and which had the most to gain from its removal. With its passing, the upward mobility so long denied to African miners would now at last be theoretically possible. Critical of their exclusion from the negotiations over the bar, and wary of what they thought were subtle state attempts at sneaking the bar back in by other means, Ramaphosa delivered a scathing attack on the proposed legislation in 1987. He claimed that the problem had begun with Wiehahn's recommendation that in the process of removing the bar the Chamber should provide employment security for those white workers most vulnerable in the labour market. This recommendation laid the basis for what became, in Ramaphosa's words, an 'inherent contradiction' in the proposed colour-blind legislation, as it sought to remove racial discrimination and yet yield 'to the demands of one particular racial group'. Nowhere in Wiehahn's and the state's considerations was there any reference at all to the 'gross iniquities that have plagued black workers as a result of job reservation'. The requirement that had been built in to allay the fears of white workers was to the NUM both 'insulting' and 'hypocritical'.

For once, the NUM agreed with the Chamber that the most desirable route of reform was simply to replace 'scheduled person' with 'competent person' in the legislation. In the third version of the bill, to be eligible for any of the certificates of competency one had to meet a range of criteria, including experience, language, health, security, age, education and training. Ramaphosa contended that this would grant officials the power to 'keep black workers out of the affected occupations in a subtle manner'. He then proceeded to take apart each one of the criteria, including those of language and education, arguing that they were disciminatory in effect. Ramaphosa contended that the motive for this requirement was 'clearly to keep the majority of black workers out of the affected occupations because they do not know the official languages'. He noted that by no fault of their own Africans have been denied access to formal schooling, and that to introduce language as a criterion was 'clearly discriminatory and is intentionally calculated at keeping black workers out of the affected occupations'.[57]

Members of the parliamentary committee failed to take the thrust of Ramaphosa's argument seriously. No acknowledgment was given to his vital point about the consequences of apartheid education on black literacy levels and language proficiency. The committee still gave more recognition to the interests of white workers, and placed their employment security above the aspirations of black workers. It was unwilling to countenance, finally, a labour market free of state intervention in the allocation of work.

From Legal Discrimination to Regulated Entry

The final regulations published in terms of the Mines and Works Amendment Act of 1988 specified that candidates for certificates of competency had to be 20 years or older, have acquired 312 qualifying shifts of work, be medically fit, and be able to communicate in Afrikaans and English, both orally and in written form. The regulations also established advisory committees to manage the issuing of each of the eleven certificates of competency. The committees had to advise the Minister about minimum educational qualifications and provide information about the demand for certificate holders in the labour market. By late 1988 the advisory committees had come into being.

Because of their controversial nature, the Department of Mineral and Energy Affairs agreed in 1990 to reconsider the regulations, and awaited the recommendations of an advisory committee made up of representatives from the NUM, MWU and the Chamber. The NUM insisted that all literate team leaders with suitable experience should be eligible for the certificates, while the Chamber did not want to be restricted to team leaders only, nor pay for any extra training expenses involved. At the start of the 1990s, therefore, the political framework in terms of which the colour bar was to be ended remained unresolved.

The end of the formal colour bar in employment was in fact accompanied by a slow and partial movement of African employees into positions historically denied them. At the end of 1988, Anglo American had graduated 18 African blasting certificate holders, and Genmin 23. By the end of 1989 the Anglo American figure had increased to 75, with 40 more workers in training. This meant that of the 165 workers who were in possession of blasting certificates at Anglo American mines, 90 were white and 75 African.

Since Job Reservation Determination no. 27 was withdrawn in 1983, Africans have become officials in sampling, surveying and ventilation operations. By the end of 1984, 368 Africans out of 2 314 were registered as officials, and by the end of 1988 this had grown to 455 out of 3 317, or 20 per cent of the mine officials. African apprentices were first indentured in 1984. At the end of 1984, some 2,3 per cent of all apprentices were African, and by 1988 this had risen to 7,4 per cent. Among artisans, 1,2 per cent of a total of 9 699 were African by 1988.

The upward mobility of African workers in the occupational structure of the mines is now theoretically possible but is limited by vested interests of state officials, mine management and the established trade unions. In effect, only when there are no whites who can fill vacancies in the affected jobs will Africans be promoted. In contrast to Zambia and Zimbabwe, where many expatriate whites left their jobs after independence, most whites in the South African mines are unlikely to leave theirs.[58] An overwhelming number of white employees whose jobs are the first in line for African promotions are Afrikaners, while many of the English-speaking employees are not expatriates, and would not leave South Africa once a more representative political framework is finally installed.

As yet the mines have not adopted formal African advancement programmes, nor considered more radical Africanisation policies. But the pressures to pursue corrective measures in employment patterns will undoubtedly increase in the light of the political changes begun in February 1990. In particular, newly empowered African constituencies will very likely make demands for a level of occupational advancement denied them under apartheid. Given the trends analysed in this chapter, there are structural obstacles in the way of the advancement of Africans that remain even after the abolition of the colour bar. Rather than providing an open sesame, the end of apartheid in employment marks simply the beginning of a much more difficult process involved in the deracialisation of work and the promotion of equality of employment opportunities.

9
Conclusion:
Social Change in a
Labour-Repressive System

In the 1980s African miners gained important rights and acquired significant power in the gold industry's labour framework. They unionised and began to participate in the creation of a bureaucratically centred system of collective bargaining. The colour bar, which had restricted African miners to unskilled and semi-skilled work, was also lifted in the course of the decade, partially as a result of the pressure their accumulated skilling brought to bear on the racial division of labour. Other gains in the areas of health, safety and conditions of service accompanied these broader institutional changes in the labour framework.

Despite the changes, migrant labour and traditional forms of hostel accommodation remained enduring realities for African workers. While the lifting of the colour bar elevated some African employees into supervisory and managerial positions, the overwhelming proportion of senior jobs were still dominated by whites, and African workers remained largely locked into manual labouring work. As for relations between white and black workers these were still wedded to paternalistic forms of racial interaction. Social change in the labour framework thus stopped short of ending migrant labour, hostels and racial paternalism in the mines. While there have been tangible and important differences made to the lives of African workers, they clearly have been partial in character.

The restricted nature of all these changes raises some fundamental theoretical questions about the limits and possibilities of further change, and about the relation between worker struggles against labour repression in the workplace and the broader struggle over state power in South Africa. Though the experience of African workers in the mines is context-specific, it also relates to broader questions about power and racial inequality in South Africa. The character of social change in the mining industry seems to contain compellingly similar themes about the limits of change in South Africa more generally.

Limits of Production Struggles

Class struggles between African miners and mine management have been effective in modifying power relations and enhancing the capacities of subordinate African groups in production. The emergence of an independent trade union, and the transition from a paternalistic system of industrial relations to bureaucratically organised forms of collective bargaining, are probably the most visible aspects of these processes. By itself the ability to organise strikes in an industry spread over 300 miles, involving up to 43 gold mines and 500 000 workers, is testimony to the increasing class capacities of African workers.

The emergent class power of African workers has been limited, however, by the labour market and the wider system of state and inter-state relations of which they are part. For one thing, production struggles did not increase the real wages of African workers in the 1980s, in spite of the development of African unionism. When African workers went on strike in the 1980s, management responded by dismissing workers *en masse* and replacing them with easily available alternative workers. In this way a flooded labour market could reduce the bargaining power of organised labour.

The NUM was even more constrained when it came to wider concerns in the labour framework. Like other African unions, the NUM's members had grievances that went beyond the workplace. The NUM declared 'war' on migrant labour and the mine hostels. It threatened to call its members out on strike if the mining houses failed to make a firm commitment to end migrant labour. While the NUM's stance made a difference in the emergent housing policies of some of the mining houses, it was unable, in the face of a weak bargaining position in production, to push the mines away from migrant labour and hostel accommodation. In sum, production struggles had an uneven impact on the labour framework, failing to erode the system of migrant labour and mine hostels.

Constructing Socialism

The NUM's longer-term answer to the problems of labour repression, and more broadly the issues of racial inequality and disparities in power relations, is socialism. An NUM political policy document of 1987 noted that 'the workers in this country are not only striving for better working conditions in the mines, but for a democratic socialist society controlled by the working class'.[1] The union's construction of a socialist future has two related aspects. The first is the struggle for workers' control over production and labour processes. The union believes that shop-floor organisation has to be strengthened, in order to curtail the arbitrary powers of management. Workers are encouraged at the same time to extend their capacities to regulate the labour processes of production. The development of more democratic forms of work organisation are held up as the route by which the power and interests of workers can be advanced, and the erosion of labour-repressive and other coercive practices effected.

The second aspect of socialism has to do with the class character of state power and state intervention in the economy. The NUM advocates that working-class interests should have a strong presence in new forms of state organisation. Through the African National Congress and the South African Communist Party with which the NUM (through COSATU) has established a formal political alliance, it has pressed for state interventions which would bolster worker interests in production, check the power of management, and drive production away from the single-minded pursuit of profit-maximisation.

These two aspects of a socialist future – workers' control over production and a strong worker presence in the state – are, for the NUM, inseparable dimensions of the same desired processes. Workers' control without appropriate state interventions to check the power of management would be fragile and weak, and therefore reproduced only with difficulty. However, through the use of state power directed by working-class constituencies in the ruling bloc, worker interests would be entrenched against those of management. On the other hand, state interventions without workers' control could subvert the interests of workers, and strengthen rather than weaken the relation between state and capital. Indeed, socialist state interventions in production without workers' control could arguably pave the way for authoritarian and undemocratic forms of industrial planning associated with centrally planned command economies. The NUM's construction of socialism insists, therefore, on extending worker-based interests both in the workplace and in the state.

In an important article, Robert Davies has stressed the importance of both emphases in the NUM's approach.[2] He argues that in order to promote workers' control the mines have to be socialised, not just nationalised: 'if nationalisation is to be part of a broader process of socialisation it needs to be accompanied by concrete changes in the organisation of labour processes and decision-making at enterprise level, which permit the working masses themselves to progressively gain control over the means of production'.[3]

Davies proceeds to argue against premature 'defensive' nationalisation, citing the example of Mozambique to make a series of compelling points about its potentially negative consequences. But his analysis, and the NUM's construction of a socialist future which it arguably supports, leave a number of questions unanswered. How far can workers' control go under circumstances where class relations and ownership patterns are not being reconstructed? In which ways can it be argued that the new state will be a workers' state, or effectively and primarily represent worker-based interests? What evidence is there to suggest that organisations representing labour (specifically the COSATU federation) will be hegemonic in the ANC–SACP–COSATU alliance? In other words, how convincing are the socialist propositions about social change in the labour-repressive framework, given the possibilities and limits outlined in this study? This question becomes particularly apposite in the light of the decline of socialist states in Eastern Europe, and the widespread unease about the veracity of the socialist discourse in genuinely promoting the interests of the working classes.

Workers' Control

Michael Burawoy has argued that the labour process in colonial racial orders reproduces 'despotic' patterns of interaction between white and black. He (and others) cite South African mining as an example of 'racial despotism' in the workplace refracting relations of racial domination in the wider state and society.[4] Indeed, one of the tasks of the NUM is to erode 'racial despotism' in the workplace, by pushing for greater 'workers' control', challenging the arbitrary powers of management, and exposing the paternalism and racism of white employees.

1987 was the year in which the NUM began its campaign for greater workers' control over productive and reproductive processes. It was hailed as the year in which workers would 'take control' over their lives. In the build-up to the strike of that year, workers were encouraged to confront the arbitrary powers of management at the workplace, subvert authority hierarchies, roll back managerial regulation of the labour process, actively expose widespread racism, and take control over mine hostels. The 1987 miners' strike was seen as a culmination of worker initiatives in these respects, and a potent demonstration of the potential of workers to seize control.

As this book has indicated, the arbitrary powers of management were in the 1980s constantly under pressure in the areas of mass dismissals and health and safety. African workers and the campaigns of the NUM, assisted by certain state bodies (such as the Industrial Court), managed to roll back managerial prerogative in some areas. This did not mean that workers now assumed control over these issues, but that their power to influence decisions and to keep management to account was much greater. Workers' control in practice came to mean a shift in the balance of power with management, in some areas of the working experience, and inconsistently so.

Only in the case of the mine hostels did worker groups manage to gain real control in 1987 during the August strike. In Chapter 7, I indicated how worker-groups and NUM strike committees took over some hostels, using their power to maintain the strike by regulating the behaviour of hostel residents, by policing their movement, and by imposing sanctions against those who broke prevailing patterns of solidarity. Yet while the NUM committees at some mines used the hostel effectively as an instrument of its emergent power, to sustain a strike that otherwise might not have lasted as long as it did, their control over hostels proved temporary. In the aftermath of the strike, management regained the upper hand and, in the case of Anglo American mines where workers' control had been well developed, frustrated the NUM's efforts to use hostels for union affairs. In the light of experience, new schemes of co-operative governance over hostels came to be voiced by management and the NUM alike, and practical proposals were subsequently discussed.[5]

During the late 1980s the workers confronted some of South Africa's most powerful industrial companies, under circumstances in which their own power in the workplace was weak. As this study has indicated, their weakness in production was reinforced by the character of wider labour markets and the nature of sub-continental inter-state relations. On their own, the workers have not been able to roll back the power of management, and while the NUM has constantly urged workers to take control, there was little opportunity to make this a viable proposition in the labour process.

Faith in the State

The NUM actively participated in processes that were intended to weaken the apartheid state, and force it to collapse. Support was publicly given to ending the various states of emergency in operation against opposition movements between 1985 and 1990. Sanctions, trade boycotts, disinvestment and divestment were underwritten because they raised the costs to the state of maintaining apartheid. In other arenas, too, the NUM participated in political campaigns challenging state authority, including boycotts, mass stay-aways, sympathy strikes, and protest actions.

Like many other African unions and political organisations, the NUM has a strong attraction to and faith in centralised state power to remedy racial inequality in economy and society. Arguably, as a result in part of the historical legacy of Afrikaner nationalism and the National Party, which used state intervention on a large scale to promote the partial interests of a white (and specifically Afrikaner) minority, African constituencies have also turned to the state as an answer to their socio-economic subordination. Despite the deep differences between black and white political constituencies, there is a common attachment to centralised state power in the prevailing political culture. Since the South African state has long held control over public resources such as housing and education, it is not surprising that workers turn to the state and not to the economy for a remedy to inequality.

The faith of the NUM in the state is driven by a largely instrumentalist conception of state power. It is believed that the present (apartheid) state consists of a set of institutions directed by dominant classes and groups in civil society, and that state officials have no durable interests of their own. It is further assumed that once a different constellation of class interests is represented in a new state, where the interests of workers should be made paramount, state officials will come under considerable compulsion to pursue policies serving the needs and demands of workers. In these terms, a workers' state would come into being when the organisations representing working-class interests occupy a 'hegemonic' position in the political alliance dominant in the state.

There is now a large literature in sociology that emphasises the autonomy and durability of state interests over those of classes represented in legislatures.[6] This literature also emphasises the relative weakness of legislatures – the only area of popular representation – relative to other state institutions, such as the judiciary, administrative bureaucracies, the police and the army. Two essential points are made in the literature. The first is that state officials reproduce themselves by pursuing goals central to the state *qua* state, such as revenue-collection, legitimation and repression – functions that do not necessarily serve the interests of the constituencies represented in the legislature. The second is that state officials do not simply occupy 'empty boxes', but constitute social and political categories – some suggest even classes[7] – with interests of their own. As certain East European states have demonstrated, a workers' state can be an ideological camouflage for the interests pursued by state officials seeking privilege and power at the expense of the working masses.[8]

NUM officials argue that centralised state power should be used to remedy inequality in the mining industry, by nationalising the mining houses. But, as the literature suggests, nationalisation does not necessarily serve the interests of workers, but rather the state officials who promote it. In the example of Zambia, the nationalisation of the copper mines brought about a closer relationship between state officials and corporate executives. In the interest of accumulation in the troubled copper industry of the 1970s, state officials used state power to ban strikes, enforce a wage freeze, and subordinate trade-union interests to party and state prerogatives.[9] As a consequence of nationalisation, the post-colonial state became less, rather than more, independent of the interests of mining capital.

Union and Party

The NUM publicly aligned itself with the goals of the ANC in 1987, and adopted the Freedom Charter as the basis of its own political programme.[10] The Charter offers no concrete proposals as to how migrant labour and hostels could be abolished, but argues that 'the wealth of our country shall be restored to our people'. For the ANC this means wealth redistribution, the reorganisation of production so as to meet basic needs, and active state reconstruction of the economy. The nationalisation of mining companies is also a crucial leg of the overall approach. Presumably, once nationalised, corporate decision-making in the mines would be encouraged, even compelled, in the desired direction by state officials, and investment policies, wage philosophies and labour practices would be biased in favour of workers against management.

The ANC's Freedom Charter was drafted in the early 1950s, when nationalisation and state ownership of productive resources were, as concepts, ascendant. Three decades later, nationalisation strategies are in decline, and questioned in even unexpected circles.[11] When the ANC was able to formulate policies for public scrutiny after being unbanned in 1990, it offered, therefore, a dated and discredited policy device, that was predictably criticised by members of the business community and the small but vocal liberal public. A senior ANC spokesperson offered a rather weak defence, arguing that 'unless someone comes up with an alternative solution, the ANC has made it clear that it is going ahead with its policy of nationalising key sectors'.[12] The South African Communist Party, unbanned at the same time, held a much more cynical view, noting that it was not 'mesmerised' by nationalisation policies.

The ANC leadership, however, couldn't quite make up its collective mind whether it intended to nationalise the mines or not. In 1989 the ANC published revised policy guidelines, which committed the organisation to a 'mixed economy', leaving considerable room for a departure from nationalisation policies. However, after his dramatic release from prison in February 1990, Nelson Mandela reaffirmed the ANC's policy of nationalising the mines, banks and other large monopolistic concerns. In September 1990, the ANC issued a 'Discussion Document on Economic Policy', based on the deliberations of two workshops held in Harare, Zimbabwe, during 1990. This was a working document, sent to local and regional ANC branches for consideration and discussion, and the ANC insisted that the document should not be read as final policy. But it does contain the most comprehensive set of pronouncements on the nature of the economy and the means of its transformation the ANC has so far produced.

While ANC officials (including Mandela) still speak about nationalising the mines, the discussion document makes no mention of it. Instead, in pursuit of an economy driven by growth through redistribution, the document recommends industrial planning, 'based on strategies targeted at specific sectors'. It notes that the ANC 'would consider using fiscal policy' to 'encourage venture capital in new mines'; would consider 'the possibility of making strategic investments in mines'; and following Zimbabwe's example, it would seek to develop a 'policy of stabilising prices through the formation of cartels'.[13] It believes, furthermore, that the power of the mining houses is much too great, and that the finance conglomerates are 'an impediment' to pursuing 'an alternative strategy'. The document commits itself to exploring 'various options in respect of ownership patterns in the mining industry'. Consideration will also be given to creating an inspectorate to police mine health and safety and mining legislation, among other things. Finally, in view of the strategic importance of mining for achieving 'national developmental objectives', the ANC will consider the nature and 'extent of state intervention and ownership'.

Nationalisation is not ruled out, but its implementation can no longer be seen as a foregone conclusion. In the face of considerable extraneous pressure to abandon nationalisation policies altogether, the ANC has recently declared its willingness merely to consider the various options. For one thing, the South African business community predictably rejected nationalisation, while foreign investors, looking beyond sanctions, became anxious about the ANC's proto-socialist rhetoric, and Western governments, who recognised the centrality of the ANC to a negotiated settlement in South Africa, urged the abandoning of nationalisation policies. Political parties within South Africa have pressed the ANC to rethink its stance on nationalisation, including the National Party, the ANC's principal negotiating adversary, which has been trying to privatise state corporations.

Where does this leave the NUM? While the ANC has committed itself to 'guaranteeing organised labour a central role in the formulation and implementation of all economic policy', holding that organised labour is 'an essential component of civil society' and exists independently of 'state and political party', yet the discussion document merely makes the timid recommendation that 'attention' should paid to the demands and needs of mine workers. It urges the end of 'racist labour practices' and calls for 'improvements in living standards and conditions of work'. It says little about how the state will use its power against management; how workers' control will be reinforced and protected; and how the state will provide the political framework that guarantees the 'hegemony' of the working class and a socialist environment. In other words, there is a lack of clarity about the role of workers' interests in the ANC construction of a future economy.

The Struggle over State Power in South Africa

The struggle over labour repression in the gold mines contains more general and compelling themes about the struggle in South Africa over state power. The ascendance and emergent power of African workers in the labour framework and the centrality of their initiatives in shaping South Africa's modern history can be seen as a part of a larger picture of change and transformation. Moreover, the social character of the limits and impediments to change in the labour framework, highlighted repeatedly in this study, is also of relevance to understanding broader patterns of political transformation in South Africa as a whole.

As in the mines, the ascending power of black African people in civil society eroded the capacity of the state to maintain apartheid practices. The campaigns and struggles of the 1970s and 1980s, the ungovernability of the townships, the manifest inability of state structures to resolve African grievances, overlaid by the mounting costs of international sanctions, pushed the state towards making fairly dramatic reforms. Under F. W. de Klerk's guidance, the traditional political framework associated with apartheid began to change. In February 1990 De Klerk lifted the ban on all previously outlawed organisations, including that imposed on its major foe, the African National Congress. Political prisoners were released, most dramatically Nelson Mandela. A process whereby apartheid legislation could be repealed was begun. State of emergency measures were relaxed. The government rapidly came closer to the point where all preconditions for a negotiated settlement and the lifting of international sanctions could presumably be met.

State officials approached the bargaining process with considerable strength and a visible confidence. Challenged consistently by the mass mobilisation campaigns in the 1980s, the state's repressive organisations had, nevertheless, remained largely cohesive and effective. The army had not lost its organisational and technological competence. The police, apparently divided along political lines and under constant pressure for alleged human rights abuses, remained a coherent force. In other areas, state officials maintained a reasonable degree of administrative efficiency. These were not state organisations under threat of disintegration. It was a state under siege and subject to extraordinary and unrelenting domestic and international challenge. But this is not the same thing as an unravelling *ancien régime* with the will and capacity to govern lost as a result of subordinate challenge.

Domestically, the De Klerk reforms received strong support from the business community, including the powerful mining houses, which see his gradualist vision of reform as the start of a transition towards an apartheid-free, capitalist-oriented society. The erosion of international sanctions reinforced the direction of the reforms, and served to add to De Klerk's growing power and confidence as his government came closer to the negotiation processes. The National Party, for decades the party of Afrikaners, now presented itself as a party of the middle and upper classes, having also opened its membership for the first time in principle to people who were not white. Although it might seem slow to South Africans, in barely one year the De Klerk government had positioned itself in favour of reform, change, negotiation, the end of apartheid, and the construction of a new political order.

As Karl van Holdt has argued, because state organisations were not disintegrating, they were not therefore vulnerable to seizure by mobilised insurrectionary groups. Because the state was 'powerful, large, wealthy and complex', and 'highly developed ... with great financial, technical and military resources', insurrectionary and revolutionary challenges were likely to fail.[14] Van Holdt proceeded to assert that the seizure of state power is therefore not on the immediate agenda, that the best the opposition can hope for is a negotiated settlement under mutually agreed-upon conditions, and that the oppositional forces must therefore entrench their emergent power in the institutions of civil society and state, reinforcing thereby their political capacities in an emergent 'war of position' with government.

The recently unbanned political organisations, specifically the ANC, had to begin the transition from being an underground movement operating from exile to a public political party. Domestic offices had to be established, and membership campaigns launched. New policies dealing with the critical aspects of the restructuring of state, economy and society were in need of formulation. Negotiations with government regarding returning exiles and releasing political prisoners had to be concluded, and mechanisms put in place for their integration in society. The magnitude and complexity of the change from being an underground organisation to becoming a political party, and the cultural and ideological shift that must accompany this transition, should not be underestimated.

The ANC has been slow to make the changes. It lacks effective public visibility. Local branches have been established, but membership drives have been disappointing. Local branches complain that they are not being consulted, and that decision-making power appears to be concentrated at head office. The administrative infrastructure is unevenly developed. Policy still tends to operate on the level of rhetoric and broad principle. Caught unprepared by the De Klerk reforms, the ANC displays all the characteristics of an organisation having to make rapid changes in a demanding and challenging environment. It has not as yet developed a political machine strong and competent enough to participate effectively in state management.

In view of the relatively weak position of the ANC in the unfolding politics of negotiations in South Africa, there is some doubt whether the ANC can do the things it says it wants to do. Although it might want to nationalise the mines, the process of negotiation will mean that it has to strike deals and make compromises with a powerful and cohesive adversary, including perhaps abandoning the goals of nationalisation and those policies designed to advance the interests of workers. While the ANC would want to alter ownership patterns, even reconstruct class relations and class structures, it might not be in the position to do so.

Processes of negotiation over the reorganisation of state structures are the least likely route by which classes and ownership patterns can be reconstructed. Typically, ownership patterns tend to be altered radically only after revolutionary situations come about, when both traditional state authority and economic organisations have largely collapsed or disintegrated, and when state power is seized or captured by insurrectionary means. In such circumstances it is possible to displace previously dominant classes, and to use newly acquired state power to advantage subordinate classes by generating new opportunities for ownership, wealth and privilege. Revolutionary situations historically provide the most propitious opportunities for the radical reconstruction of class relations.[15]

On the other hand, negotiation politics provide the most unlikely circumstances for the transformation of ownership patterns and the reconstruction of class relations. Inasmuch as the political capacities of the ANC are still weakly developed in the face of cohesive dominant classes and state, the chances of putting new classes in place of old are remote. Negotiation politics are likely to reproduce existing power and class relations, rather than transform them. They involve changing the racial character of the state, and not the class structures of society at large.

Like the NUM in the labour framework, the ANC has made a number of important gains and advances in the emergent political framework. It is, as yet, a young and maturing organisation, and needs time to develop an effective infrastructure to service the interests of its members. However, while the ANC intends to introduce state intervention on a large scale, it might be obliged to settle for a great deal less. As a result of the negotiated character of change, the ability of a reconstructed state to transform class relations and ownership patterns in the political economy will be seriously limited.[16] The capacity of a reconstructed state to intervene effectively in production processes on behalf of workers' constituencies and unions is likewise doubtful. Like the NUM in the labour framework, the ANC can advance the power of black people in state and society, without, however, conquering state or society.

References

Chapter 1

1. I adhere to the convention of referring to Bantu-speaking people as Africans or black Africans; 'blacks' is an inclusive term for African, Indian and coloured people.
2. The reconstruction is based on Sheila van der Horst, *Native Labour in South Africa* (Cape Town, 1942), pp. 66–83, 110–18, 157–72, 186–232; Alan Jeeves, *Migrant Labor in South Africa's Mining Economy* (Kingston and Montreal, 1985), pp. 37–86; Stanley B. Greenberg, *Race and State in Capitalist Development: Comparative Perspectives* (New Haven, 1980), pp. 148–75; Francis Wilson, *Labour in the South African Gold Mines 1911–1969* (Cambridge, 1972), pp. 1–44; and Charles van Onselen, *Studies in the Social and Economic History of the Witwatersrand. 1: New Babylon* (London and Johannesburg, 1982), pp. 1–43.
3. Colin Bundy, *The Rise and Fall of the South African Peasantry* (London, 1979), pp. 109–45, 197–220; Wilson, *Labour in the South African Gold Mines*, pp. 2–3; Van der Horst, *Native Labour in South Africa*, pp. 298–318.
4. Frederick Johnstone, *Class, Race and Gold: A Study of Class Relations and Racial Discrimination in South Africa* (London, 1976), pp. 42–5; Greenberg, *Race and State in Capitalist Development*, p. 162.
5. Wilson, *Labour in the South African Gold Mines*, pp. 54–5.
6. Jeeves, *Migrant Labor in South Africa's Mining Economy*, pp. 121–86.
7. Wilson, *Labour in the South African Gold Mines*, p. 5.
8. Greenberg, *Race and State in Capitalist Development*, pp. 165–7.
9. Jeeves, *Migrant Labor in South Africa's Mining Economy*, pp. 37–58.
10. David Yudelman and Alan Jeeves, 'The closing of the frontier: Black migrants to the South African mines, 1920–1985', *Journal of Southern African Studies*, 13, no. 1 (1986), pp. 101–24. See also Jonathan Crush, 'The extrusion of foreign labour from the South African gold mining industry', *Geoforum*, 17, no. 2 (1986), pp. 161–72; Jonathan Crush, Alan Jeeves and David Yudelman, *South Africa's Labor Empire: A History of Black Mineworkers under Apartheid* (Boulder, forthcoming).
11. William Worger, *South Africa's City of Diamonds: Mine Workers and Monopoly Capitalism in Kimberley 1867–1895* (New Haven, 1987), pp. 141–4.
12. Charles van Onselen, *Chibaro: African Mine Labour in Southern Rhodesia, 1900–1933* (London, 1976), pp. 136–8; Johnstone, *Class, Race and Gold*, pp. 38–9; Sean Moroney, 'The compound as a mechanism of worker control 1900–1912', *South African Labour Bulletin*, 4, no. 3 (1978), pp. 29–49; John Rex, 'The compound, reserve and urban location', *South African Labour Bulletin*, 1, no. 4 (1974), pp. 4–17.

13. Johnstone, *Class, Race and Gold*, pp. 168–200.

14. Dan O'Meara, 'The 1946 African mineworkers' strike and the political economy of South Africa', *Journal of Commonwealth and Comparative Politics*, 13, no. 2 (1975), pp. 146–71; Dunbar Moodie, 'The moral economy of the black miners' strike of 1946', *Journal of Southern African Studies*, 13, no. 1 (1986), pp. 1–35; Wilmot G. James, 'Grounds for a strike: South African gold mining in the 1940s', *African Economic History*, no. 17 (1987), pp. 1–22.

15. Dunbar Moodie, 'The formal and informal social structure of a South African gold mine', *Human Relations*, 33, no. 8 (1980), pp. 555–74.

16. The concept 'labour repression' was first introduced to South African studies by Stanley Trapido, and later developed by Stanley B. Greenberg. See Trapido, 'South Africa in a comparative study of industrialization', *Journal of Development Studies*, 7, no. 3 (1971), pp. 309–20; and Greenberg, *Race and State in Capitalist Development*, pp. 54–69, 70–86. A useful review of the 'labour-repressive' literature can be found in Jonathan Hyslop, 'A Prussian path to apartheid? Germany as comparative perspective in critical analysis of South African society', *South African Sociological Review*, 3, no. 1 (1990), pp. 33–55.

17. Republic of South Africa, *Report of the Commission of Inquiry into Legislation Affecting the Utilization of Manpower* (Riekert), (Pretoria, RP32/1979); *Report of the Commission of Inquiry into Labour Legislation* (Wiehahn), (Pretoria, RP47/1979, 38/1980, 32/1980, 87/1980, 27/1981, 28/1981).

18. The Employment Bureau of Africa (TEBA), 'Black labour force: Need for the industry to maintain the right to select its own labour force' (Johannesburg, 1984).

19. See Merle Lipton, 'Men of two worlds: Migrant labour in South Africa', *Optima*, 29, nos.2 and 3 (1980), pp. 72–201.

20. Steven Friedman, *Building Tomorrow Today: African Workers in Trade Unions 1970–1984* (Johannesburg, 1987), pp. 355–62.

21. Jonathan Crush, 'Migrancy and militance: The case of the National Union of Mineworkers of South Africa', *African Affairs*, 88, no. 350 (1989), pp. 15–23.

22. Marcel Golding, 'Mass dismissals on the mines: The workers' story', *South African Labour Bulletin*, 10, no. 7 (1985), pp. 77–117.

23. TEBA, 'Summary of Monthly Field Reports' ('SMR'), (4th Quarter 1987).

24. Harold Wolpe, 'Capitalism and cheap labour power in South Africa: From segregation to apartheid', *Economy and Society*, 1, no. 4 (1972), pp. 425–56.

25. For a useful review of this literature see Robin Cohen, *The New Helots: Migrants in the International Division of Labour* (Aldershot, 1988), pp. 73–110. Important criticisms of Wolpe's arguments can be found in Charles Simkins, 'Agricultural production in the African reserves of South Africa, 1911–1969', *Journal of Southern African Studies*, 7, no. 2 (1981), pp. 256–83.

26. Michael Burawoy, 'The functions and reproduction of migrant labor: Comparative material from Southern Africa and the United States', *American Journal of Sociology*, 81, no. 5 (1977), pp. 1 051–87.

27. For a similar point see also Ruth First, *Black Gold: The Mozambican Miner, Peasant or Proletarian* (Brighton, 1983), p. 8.

28. Burawoy is aware of the functionalist tendencies of reproduction analysis. In a footnote he writes that 'throughout ... I associate the state with the organization of the reproduction of systems of migrant labor, thus unavoidably conferring upon it a monolithic quality it does not in practice possess'. Burawoy, 'The functions and reproduction of migrant labor', p. 1 053.

29. Greenberg, *Race and State in Capitalist Development*.

30. Stanley B. Greenberg, *Legitimating the Illegitimate: State, Markets and Resistance in South Africa* (Berkeley, 1987).

31. David Yudelman, *The Emergence of Modern South Africa: State, Capital and the Incorporation of Organized Labor on the South African Gold Fields, 1902–1939* (Westport and Cape Town, 1983).

32. Merle Lipton, *Capitalism and Apartheid: South Africa 1910–1984* (Aldershot and Cape Town, 1986).

33. See Robert Davies, David Kaplan, Mike Morris and Dan O'Meara, 'Class struggle and the periodisation of the South African state', *Review of African Political Economy*, 7 (1978), pp. 4–30. Dan O'Meara's study of the African miners' strike of 1946, while purporting to be about African workers, is more interested in the impact of the strike on dominant groups and the state. See O'Meara, 'The 1946 African miners' strike and the political economy of South Africa'.

34. In this regard, the *South African Labour Bulletin* has played a key role. See Johann Maree, ed., *The Independent Trade Union Movement: Ten Years of the Labour Bulletin* (Johannesburg, 1989); Jon Lewis, *Industrialization and Trade Union Organization in South Africa, 1924–1955: The Rise and Fall of the South African Trades and Labour Council* (Cambridge, 1984); Eddie Webster, *Cast in a Racial Mould: Labour Process and Trade Unionism in the Foundries* (Johannesburg, 1987); Eddie Webster, ed., *Essays in Southern African Labour History* (Johannesburg, 1978).

35. Ari Sitas, 'Moral formations and struggles amongst migrant workers on the East Rand', *Labour, Capital and Society*, 18, no. 2 (1985), pp. 372–401; Moodie, 'The formal and informal structure of a gold mine'; J. K. McNamara, 'Inter-group violence among black employees on South African gold mines', *South African Sociological Review*, 1, no. 1 (1988), pp. 23–38.

36. Jean Leger and Philip van Niekerk, 'Organizing on the mines: The NUM phenomenon', in *South African Review 3* (Johannesburg, 1985), pp. 68–73; Friedman, *Building Tomorrow Today*, pp. 355–92; Clive Thompson, 'Black trade unions on the mines', *South African Review 2* (Johannesburg, 1984), pp. 156–164; C. Pycroft and B. Munslow, 'Black mine workers in South Africa: Strategies of co-option and resistance', *Journal of Asian and African Studies*, 23, nos. 1 and 2 (1988), pp. 156–79; Crush, 'Migrancy and militance: The case of the National Union of Mineworkers of South Africa'.

37. R. Rafel, 'Job reservation on the mines', in *South African Review 4*, eds. G. Moss and I. Obery (Johannesburg, 1987), pp. 265–82; Jeff Lever and Wilmot G. James, 'Towards a deracialised labour force: Industrial relations and the abolition of the job colour bar on the South African gold mines' (Stellenbosch, 1987); Jeff Lever, 'Established trade unions and industrial relations on the gold mines in the 1980s', *Industrial Relations Journal of South Africa*, 8, no. 2 (1988), pp. 1–14.

38. Jon Lewis, 'South African labour history: An historiographical assessment', *Radical History Review*, 46/7 (Winter 1990), pp. 213–35.

39. Rob Lambert, 'Trade unions, nationalism, and the socialist project in South Africa', in *South African Review 4*, eds. G. Moss and I. Obery (Johannesburg, 1987), pp. 232–52.

40. Michael Burawoy, *The Politics of Production: Factory Regimes under Capitalism and Socialism* (London, 1985).

41. In a footnote Burawoy writes: 'This is not to say that political relations among states are not important but rather that they assume less significance with the consolida-

tion of the capitalist mode of production in peripheral social formations.' Burawoy, 'The hidden abode of underdevelopment: Labor process and the state in Zambia', *Politics and Society*, 11, no. 2 (1982), p. 130.

42. Charles Ragin, *The Comparative Method: Moving Beyond Qualitative and Quantitative Strategies* (Berkeley, 1987), pp. 34–52.

43. Ragin, *The Comparative Method*, pp. 2–3, 34–52.

44. Van der Horst, *Native Labour in South Africa*; Jeeves, *Migrant Labor in South Africa's Mining Economy*; Norman Levy, *The Foundations of the South African Cheap Labour System* (London, 1982); Van Onselen, *Studies in the Social and Economic History of the Witwatersrand*; Johnstone, *Class, Race and Gold*; Worger, *South Africa's City of Diamonds*; Robert Turrell, *Capital and Labour on the Kimberley Diamond Fields 1871–1890* (Cambridge, 1987); Robert Davies, *Capital, State and White Labour in South Africa 1900–1960* (Brighton, 1979).

45. H. J. and R. E. Simons, *Class and Colour in South Africa 1850–1950* (Harmondsworth, 1968), pp. 271–99; Davies, *Capital, State and White Labour*; Yudelman, *The Emergence of Modern South Africa*, pp. 164–89; O'Meara, 'The 1946 African mineworkers' strike'; Moodie, 'The moral economy of the 1946 black miners' strike'; James, 'Grounds for a strike'.

46. Fion de Vletter, 'Recent trends and prospects of black migration to South Africa', *Journal of Modern African Studies*, 23, no. 4 (1985), pp. 667–702; Fion de Vletter, 'Foreign labour on the South African gold mines: New insights into an old problem', *International Labour Review*, 126, no. 2 (1987), pp. 199–218; Crush, 'The extrusion of foreign labour'; Wilmot G. James, 'Politics and economics of internalisation; Labour migracy to the South African gold mines 1980–2000' (Cape Town, 1987); Leger and Van Niekerk, 'Organizing on the mines: The NUM phenomenon'; Friedman, *Building Tomorrow Today*, pp. 355–92; Thompson,'Black trade unions on the mines'; Pycroft and Munslow, 'Black mine workers in South Africa'; Jean Leger, 'Safety and the organisation of work in South African gold mines: A crisis of control', *International Labour Review*, 125, no. 5 (1986), pp. 591–603; Rafel, 'Job reservation on the mines'; Lever and James, 'Towards a deracialised labour force'; Lever, 'Established trade unions and industrial relations on the gold mines in the 1980s'.

47. Earlier general works that end in the 1940s and 1950s, but constitute essential reading, are Simons and Simons, *Class and Colour in South Africa*, and Eddie Roux, *Time Longer than Rope* (Madison, 1964); see 'Implications', in Wilson, *Labour in the South African Gold Mines*, pp. 140–55; Greenberg, *Race and State in Capitalist Development*, pp. 385–410.

48. Yudelman, *The Emergence of Modern South Africa*, pp. 263–90.

49. Lipton, *Capitalism and Apartheid*.

50. See Philip Abrams, *Historical Sociology* (Ithaca, 1982); and Theda Skocpol, ed., *Vision and Method in Historical Sociology* (New York, 1984).

Chapter 2

1. Chamber of Mines, *Annual Report* (Johannesburg, 1987), p. 15S. The Soviet estimate is from *Gold* (London, Consolidated Goldfields Ltd, 1988).

2. Yudelman, *The Emergence of Modern South Africa*, pp. 29–43.

3. Wilson, *Labour in the South African Gold Mines*, pp. 36–39; Johnstone, *Class, Race and Gold*, pp. 17–20; Van Onselen, *Studies in the Social and Economic History of the Witwatersrand*, Vol. 1, pp. 1–43.

4. Timothy Green, *The World of Gold* (London, 1980), pp. 107–123.

5. Yudelman, *The Emergence of Modern South Africa*, p. 267.

6. Miklos Salamon, 'Research and development in the gold mining industry' (Chamber of Mines, Memorandum, 3 June 1974).

7. Chamber of Mines Research Organization, *Annual Report* (Johannesburg, 1974–1986).

8. Duncan Innes, *Anglo American and the Rise of Modern South Africa* (Johannesburg, 1984), pp. 271–333.

9. For a detailed description of how the system works, see Wilson, *Labour in the South African Gold Mines*, pp. 26–33.

10. In his book, *Bullion Johannesburg: Men, Mines and the Challenge of Conflict* (Johannesburg, 1986), John Lang gives an account of the internal dynamics of the Chamber based for the very first time on his access to its extensive but normally sequestered archives. Lang provides a fascinating and revealing narrative about the personalities and workings of the Chamber, in the context of its broader functions in the production and marketing of gold. He covers the history of the Chamber, from its formation in 1889. Unfortunately, Lang underplays the divisions and conflicts within the Chamber in the 1970s and 1980s, when the labour framework underwent major changes. Disagreements within the Chamber about the new industrial relations framework are largely swept under the carpet, portrayed as if they were minor aberrations in an otherwise unified Chamber. Lang, therefore, promotes the myth of the cohesive Chamber.

11. Transvaal Chamber of Mines, *Gold in South Africa*, p. 10B, quoted by Wilson, *Labour in the South African Gold Mines*, p. 20.

12. *Sunday Tribune*, 2 April 1989, p. 21.

13. For a description of the labour process see Michael O'Donovan, 'The labour process in gold mining' (University of the Witwatersrand: unpublished Honours dissertation, 1985).

14. Wilson, *Labour in the South African Gold Mines*, pp. 20–1.

15. Jean Leger, 'Key issues in safety and health in South African mines', *South African Sociological Review*, 2, no. 2 (1990), p. 1.

16. *Sunday Tribune*, 2 April 1989, p. 21.

17. David Frost, 'The political economy of trackless mining (TM3) in the South African gold mines' (Cape Town, unpublished paper, 1988).

18. Technical Assistance Group (TAG), *New Technologies on the Mines* (Johannesburg, 1986); David Frost, 'Hesistant revolution: Mechanisation, labour and the gold mines', (University of Cape Town: unpublished Honours thesis, 1987); Frost, 'The political economy of trackless mining'.

19. Frost, 'The political economy of trackless mining', p. 11.

20. The distinction is taken from Burawoy, 'The hidden abode of underdevelopment: Labour process and the state in Zambia', pp. 146–7.

21. Jonathan Crush, 'Restructuring migrant labour on the gold mines', in *South African Review 4*, eds. G. Moss and I. Obery (Johannesburg, 1987), pp. 283–91.

22. Crush, 'Restructuring migrant labour', p. 286; TEBA, *Summary of Monthly Reports* (4th Quarter 1988).

23. TEBA, *Annual Report and Financial Statement* (Johannesburg, 1976–1988).

Chapter 3

1. Yudelman and Jeeves, 'New labour frontiers for old: Black migrants to the South

African gold mines', p. 117.

2. First, *Black Gold*, pp. 3–4, 58–9.

3. The literature on labour-sourcing assumed that the Chamber either took a decision to internalise the labour force or condoned it after the fact. Sources consulted for this study confirmed that it was a deliberate decision. TEBA, 'Black labour force: Need for the industry to maintain the right to select its own labour force' (Johannesburg, 1984), p. 3.

4. TEBA, 'Black labour force', p. 2.

5. TEBA, 'Teba strategic plan 1987–2000' (Johannesburg, 1987).

6. James Cobbe, 'Consequences for Lesotho of changing South African labour demand', *African Affairs*, 85, no. 338 (January 1986), pp. 23–48; Colin Murray, 'Migrant labour and changing family in the rural periphery of southern Africa', *Journal of Southern African Studies*, 6, no. 2 (April 1980), pp. 139–156; Colin Murray, 'Migration, differentiation and the developmental cycle in Lesotho', *African Perspectives*, 2 (1978), pp. 127–43.

7. Jack Parson, 'The peasantariat and politics: Migration, wage labour and agriculture in Botswana', *Africa Today*, 31, no. 4 (1984), pp. 5–25; Jonathan Crush, 'Swazi migrant workers and the Witwatersrand gold mines 1886–1920', *Journal of Historical Geography*, 12, (1986), pp. 27–40; Alan Booth, 'Capitalism and the competition for Swazi labour, 1945–1960', *Journal of Southern African Studies*, 13, no. 1 (October 1986), pp. 125–50.

8. Greenberg, *Race and State in Capitalist Development*, pp. 148–75.

9. Yudelman, *The Emergence of Modern South Africa*, pp. 37–42.

10. See Davies, Kaplan, Morris and O'Meara, 'Class struggle and the periodisation of the state in South Africa', pp. 4–30.

11. In his study of Zambian labour, Michael Burawoy dismissed the importance of state interests during the period when, as he put it, the 'capitalist mode of production' in 'peripheral social formations' became dominant. Burawoy, 'The hidden abode of underdevelopment: Labor process and the state in Zambia', p. 130.

12. See John Suckling and Landeg White, eds., *After Apartheid: Renewal of the South African Economy* (London, 1988).

13. Theda Skocpol, *States and Social Revolutions* (Cambridge, 1980), pp. 19–33.

14. Michael Mann, *States, War and Capitalism: Studies in Political Sociology* (Oxford, 1988), pp. 1–72.

15. A. Whiteside and C. Patel, 'Agreements concerning the employment of foreign black labour in South Africa' (Geneva, 1985), p. 31.

16. TEBA, 'Black labour force', p. 9.

17. Whiteside and Patel, 'Agreements concerning the employment', p. 31.

18. J. K. McNamara, 'Black worker conflicts on South African gold mines' (University of the Witwatersrand: unpublished Ph.D. dissertation, 1985), p. 91.

19. TEBA, 'Black labour force', p. 9.

20. Grete Brochmann, 'Migrant labour and foreign policy: The case of Mozambique', *Journal of Peace Research*, 22, no. 4 (1985), pp. 335–44.

21. First, *Black Gold*, p. 334.

22. Minister of Manpower, to President of the South African Chamber of Mines, 6 June 1984, reproduced in TEBA, 'Black labour force'.

23. TEBA, 'Black labour force', Annexure IIA, p. 1.

24. Chamber of Mines, Members' circulars: Minutes of meeting, Chamber of Mines and Mozambique officials, Maputo, 17 January 1986.

25. TEBA, Memorandum, ACC/20/MOC/26/86, 15 April 1986; Chamber of Mines, Members' circulars: Minutes of Meeting, Chamber and Mozambique officials, 17 January 1986; Minutes of Meeting, Chamber (Technical Advisory Committee) and Mozambique officials, 28 March 1985; Chamber of Mines, 'Report of RSA/Mozambique Technical Committee in regard to labour in the mining sector' (6 February 1985, 28 March 1985).

26. TEBA: A. C. Fleischer, 'TEBA five year plan' (Johannesburg, 1982).

27. Chamber of Mines, Members' circulars: Minutes of meeting, Chamber of Mines and Mozambique officials, Maputo, 17 January 1986.

28. Jean Leger, 'Mozambican miners' reprieve', *South African Labour Bulletin*, 2, no. 2 (1987), p. 29.

29. D. MacArthur, J. B. Godfrey, K. McNamara and O. F. Thomas, *Factors Affecting the Popularity of a Mine for Black Mineworkers* (Johannesburg, COMRO Report No. 58/77, 1978).

30. TEBA, Circular letter no. TEBA 47/86, 17 December 1986.

31. Chamber of Mines, *Annual Report* (1988), p. 39S.

32. See Robert E. Christiaansen and Jonathan G. Kydd, 'The return of Malawian labour from South Africa and Zimbabwe', *Journal of Modern African Studies*, 21, no. 2 (1983), pp. 311–26.

33. TEBA: Fleischer, 'TEBA five year plan'; Chamber of Mines, circular Y.67/82; Chamber of Mines, Members' circulars: Note for the Record (c. January 1983).

34. TEBA: Fleischer, 'TEBA five year plan'.

35. On AIDS in Africa, see Dieter Koch-Weser and Hennelore Vanderschmidt, eds., *The Heterosexual Transmission of AIDS in Africa* (Cambridge, 1988); Norman Miller and Richard Rockwell, eds., *AIDS in Africa: The Social and Policy Impact* (Lewiston, NY, 1988); Richard Chirimuuta, *Aids, Africa and Racism* (Derbyshire, 1987); and Bekki Johnson, *AIDS in Africa: A Review of Medical, Public Health, Social Science and Popular Literature* (Aachen, West Germany, 1988).

36. TEBA, 'SMR' (Malawi, 2nd Quarter 1986).

37. TEBA, 'SMR' (Malawi, 2nd Quarter 1986).

38. TEBA, 'SMR' (Malawi, February 1987).

39. TEBA, 'SMR' (Botswana, April 1986; Lesotho, May 1987).

40. SAIRR, *Race Relations Survey 1987/1988* (Johannesburg, 1988), p. 316.

41. Interestingly, this was not the first time that Malawian workers were phased out of the migrant labour system as a result of disease. In the 1910s and 1920s, Malawian workers were refused employment because of their susceptibility to tuberculosis. See Randall Packard, *White Plague, Black Labor: Tuberculosis and the Political Economy of Health and Disease in South Africa* (Pietermaritzburg, Berkeley and London, 1990); Packard, 'Tuberculosis and the development of industrial health policies on the Witwatersrand, 1902–1932', *Journal of Southern African Studies*, 13, no. 2 (1987), pp. 187–209.

42. Cobbe, 'Consequences for Lesotho of changing South African labour demand', p. 37.

43. TEBA, 'SMR' (Lesotho, August 1985).

44. McNamara, 'Inter-group violence among black employees on South African gold mines', pp. 26–8.

45. TEBA: Fleischer, 'TEBA five year plan'.

46. TEBA, 'Black labour force', p. 8.

47. TEBA, Memorandum, X/BDV (1 February 1986).

48. TEBA, Memorandum, ACC.20/MOC/26/86 (15 April 1986).

49. McNamara, 'Inter-group violence among black employees on South African gold mines', p. 27.
50. TEBA, 'SMR' (Lesotho, 1st Quarter 1986).
51. TEBA, 'SMR' (Lesotho, June 1987).
52. TEBA, 'SMR' (Lesotho, August 1987).
53. TEBA, 'SMR' (Lesotho, September 1987).
54. TEBA, 'SMR' (Lesotho, October 1987).
55. TEBA, 'Black labour force', p. 8.
56. TEBA, 'SMR' (Swaziland, 4th Quarter 1986; 4th Quarter 1987).
57. TEBA, 'SMR' (Swaziland, October 1987).
58. W. R. Bohning, ed., *Black Migration to South Africa: A Selection of Policy-Oriented Research* (Geneva, 1981), p. 76.
59. Bohning, *Black Migration to South Africa*, pp. 11, 77, 149.

Chapter 4

1. Yudelman and Jeeves, 'New labour frontiers for old: Black migrants to the South African gold mines', p. 108.
2. James, 'Grounds for a strike: South African gold mining in the 1940s', pp. 2–6.
3. Yudelman and Jeeves, 'New labour frontiers for old: Black migrants to the South African gold mines', pp. 123–4.
4. TEBA, 'Black labour force', p. 13.
5. See summaries of this literature in Francis Wilson and Mamphela Ramphele, *Uprooting Poverty: The South African Challenge* (Cape Town, New York and London, 1989), pp. 69–73; and Cohen, *The New Helots*, pp. 73–110.
6. Yudelman, *The Emergence of Modern South Africa*, pp. 267–8.
7. TEBA, 'TEBA strategic plan 1988–2000' (Johannesburg, 1988).
8. Burawoy, 'The functions and reproduction of migrant labor', pp. 1 051–2..
9. See, for example, Transvaal Chamber of Mines, Gold Producers' Committee, *Statements of Evidence, Statistics and Memoranda, Submitted to the Witwatersrand Mine Wages Commission* (Johannesburg, 1943).
10. Wolpe, 'Capitalism and cheap labour-power in South Africa', pp. 440–1.
11. Simkins, 'Agricultural production in the African reserves of South Africa', pp. 256–83; and Cohen, *The New Helots*, pp. 94–110.
12. Wilson and Ramphele, *Uprooting Poverty*, pp. 62–3.
13. See Jill Nattrass and Julian May, 'Migration and dependency: Sources and levels of income in Kwazulu', *Development Southern Africa*, 3, no. 4 (1986), pp. 583–99; J. D. May, 'Migrant labour in Transkei - Cause and consequences on a village level' (University of Natal, 1985); V. Makanjee, 'Bophuthatswana: Bordering on no-man's land', *Indicator South Africa*, 5, no. 4 (1988), pp. 39–44.
14. AAC: Market Research Africa, 'Mine and accommodation study' (Johannesburg, 1988).
15. See Belinda Bozzoli, 'Marxism, feminism and South African studies', *Journal of Southern African Studies*, 9, no. 2 (1983), pp. 139–171, especially 143–5.
16. Amelia Mariotti, 'The incorporation of African women into wage employment in South Africa, 1920–70' (University of Connecticut: unpublished Ph.D. dissertation, 1980).
17. Cohen, *The New Helots*, pp. 81–2.
18. Laureen Platzky and Cherryl Walker, *The Surplus People: Forced Removals in South*

Africa (Johannesburg, 1985) pp. 128–88.

19. Charles Simkins, *Four Essays on the Past, Present and Future Distribution of the African Population of South Africa* (Cape Town, 1983), pp. 53–7. See also Simkins, 'What has been happening to income distribution and poverty in the homelands?' (University of Cape Town, Carnegie Papers no. 7, 1984).

20. Platzky and Walker, *The Surplus People*, p. 334. Unemployment figures for South Africa, and particularly the homelands, vary considerably from one estimate to another.

21. TEBA, 'Black labour force', pp. 11–12; Annexure 1, p. 1.

22. Director-General of the Department of Co-operation and Development (DCD), to Chamber of Mines, 25 August 1981, reproduced in TEBA, 'Black labour force'.

23. Director-General, DCD, to TEBA, 1 April 1982, reproduced in TEBA, 'Black labour force'.

24. Director-General, DCD, to TEBA, 1 April 1982, reproduced in TEBA, 'Black labour force'.

25. TEBA, 'Black labour force', p. 15.

26. Jeff Guy and Motlatsi Thabane, 'Technology, ethnicity and ideology: Basotho miners and shaft-sinking on the South African gold mines', *Journal of Southern African Studies*, 14, no. 2 (1988), pp. 257–78.

27. TEBA, 'Black labour force', p. 14.

28. Wilmot G. James, 'Migrant labour selection in South Africa's metal mining industry' (Kingston, Canadian Association of African Studies, 1988).

29. SAIRR, *Survey of Race Relations 1987/88* (Johannesburg, 1988), p. 61.

30. TEBA, 'SMR' (4th Quarter 1985).

31. TEBA, 'SMR' (January 1987).

32. *Indicator SA*, 4, no. 3 (1987), pp. 93–96; also J. Sharp, 'Relocation and the problem of survival in Qwa Qwa: A report from the field', *Social Dynamics*, 8, no. 2 (1982), pp. 11–29; J. Martiny and J. Sharp, 'An overview of Qwa Qwa: Town and country in a South African bantustan' (Cape Town, Carnegie conference paper no. 286, 1984); J. Sharp and A. Spiegel, 'Vulnerability to impoverishment in South Africa's rural areas' (Cape Town, Carnegie conference paper no. 52, 1984).

33. SAIRR, *Race Relations Survey 1987/88*, p. 887; SAIRR, *Race Relations Survey 1988/89* (Johannesburg, 1989), pp. 89–90.

34. TEBA, 'SMR' (February 1987).

35. TEBA, 'SMR' (3rd Quarter 1985).

36. TEBA, 'SMR' (1st Quarter 1986).

37. TEBA, 'SMR' (4th Quarter 1986).

38. See Jeeves, *Migrant Labor in South Africa's Mining Economy*.

39. TEBA, 'SMR' (4th Quarter 1985; 4th Quarter 1986; October 1987).

40. Chamber of Mines, Members' circulars: Minutes of meeting with Transkei officials (21 June 1985); TEBA, 'SMR' (Ciskei, 4th Quarter 1985).

41. Chamber of Mines, Members' circulars: Minutes of meeting with Transkei officials.

42. TEBA, 'SMR' (May 1986; July 1986).

43. TEBA, 'SMR' (May 1986).

44. See Tom Lodge, 'Political mobilisation in the 1950s: An East London case study', in *The Politics of Race, Class and Nationalism in Twentieth Century South Africa*, eds. Shula Marks and Stanley Trapido (London, 1987), pp. 310–35; Helen Bradford, *A Taste of Freedom: The ICU in Rural South Africa, 1924–30* (New Haven and Johannesburg, 1984).

45. Leger and Van Niekerk, 'Organizing on the mines: The NUM phenomenon', p. 73.
46. TEBA, 'SMR' (2nd Quarter 1988).
47. TEBA, 'SMR' (Transkei, June 1987).
48. Crush, 'Restructuring migrant labour on the gold mines', p. 284.
49. TEBA, 'SMR' (Bophuthatswana, 3rd Quarter 1986).
50. TEBA, 'SMR' (KwaZulu, March 1987).
51. TEBA, 'SMR' (Ciskei, February 1986).
52. TEBA, 'SMR' (Ciskei, 4th Quarter 1985).
53. TEBA, 'SMR' (Ciskei and Transkei, December 1986).
54. TEBA, 'SMR' (Bophuthatswana, September 1987).
55. TEBA, 'SMR' (Bophuthatswana, August 1987).
56. TEBA, 'SMR' (KwaZulu, 1st Quarter 1986).
57. TEBA, 'SMR' (Orange Free State, January 1986).
58. TEBA, 'SMR' (Ciskei, May 1986).
59. The term is taken from Greenberg, *Legitimating the Illegitimate*, p. 53.
60. TEBA, 'SMR' (Ciskei, 2nd Quarter 1986).
61. TEBA, 'SMR' (Ciskei, 4th Quarter 1985).
62. TEBA, 'SMR' (Orange Free State, 1st Quarter 1987).
63. Jeeves, *Migrant Labor in South Africa's Mining Economy*, pp. 153–8.
64. Greenberg, *Legitimating the Illegitimate*, pp. 53–5; Stanley B. Greenberg and Hermann Giliomee, 'Managing influx control from the rural end: The black homelands and the underbelly of privilege', in *Up Against the Fences: Poverty, Passes and Privilege in South Africa* (Cape Town, 1985), pp. 68–84.

Chapter 5

1. TEBA, 'Black labour force', p. 13.
2. Figures on urban recruiting were provided by Jonathan Crush.
3. W. Z. Mashaba and M. H. Steen, 'A preliminary investigation into the feasibility of attracting urbanised labour to work in mining' (Chamber of Mines, 1986), p. 13.
4. T. Dunbar Moodie, 'Migrancy and male sexuality on the South African gold mines', *Journal of Southern African Studies*, 14, no. 2 (1988), pp. 234, 241–2.
5. See Moodie, 'The formal and informal social structure of a South African gold mine', pp. 555–74; James Leatt, Paulus Zulu, Manoko Nchwe, Mark Ntshangase and Richard Laughlin, 'Reaping the whirlwind? Report on a joint study by the National Union of Mineworkers and Anglo American gold division on the causes of mine violence', edition with protocols (Johannesburg, 1986).
6. TEBA, 'Black labour force', Annexure VI, p. 2.
7. See David Brown, 'The basement of Babylon: Language and literacy on the South African gold mines', *Social Dynamics*, 14, no. 1 (1988), pp. 46–56.
8. Van Onselen, *Chibaro*, p. 152.
9. The discussion of Zimbabwean workers relies on McNamara, 'Black worker conflicts on South African gold mines', pp. 151–85; and McNamara, 'Implications of the Rhodesian labour experiment for the recruitment of South African urban township residents into the gold mines' (Chamber of Mines, 1985).
10. A point made by a senior member of management of a large mining house.
11. RMLD, Memo: Black Unrest on Blyvooruitzicht, 22–24 August, quoted by McNamara, 'Black worker conflicts', p. 160.
12. McNamara, 'Black worker conflicts', p. 159.

13. TEBA Liaison Division: Unrest Files, quoted by McNamara, 'Black worker conflicts', p. 169.
14. Chamber of Mines, 'Native labour – Rhodesia, 1976', quoted by McNamara, 'Black worker conflicts', p. 161.
15. W. Z. Mashaba and M. H. Steen, 'ERPM local recruitment experiment' (Chamber of Mines, 1986), p. 1.
16. TEBA, 'Black labour force', Annexure VI, p. 1.
17. TEBA, 'Black labour force', Annexure VI, p. 1.
18. Mashaba and Steen, 'ERPM local recruitment experiment', p. 1; TEBA, 'Aide Memoire' (24 August 1985).
19. Mashaba and Steen, 'ERPM local recruitment experiment', p. 2.
20. Figures on novice recruiting and mine reputation were supplied by J. K. McNamara.
21. Chamber of Mines, Loss Control Surveys (Chamber of Mines, 1984–1986).
22. Chamber of Mines Research Organisation, Human Resources Laboratories, 'Memorandum' (14 August 1985).
23. Mashaba and Steen, 'A preliminary investigation', p. 6; Chamber of Mines, 'Report of a meeting held on 7 January 1986, Note for the record'.
24. All subsequent quotations are from Mashaba and Steen, 'A preliminary investigation', pp. 9–14.
25. Pundy Pillay, 'Future labour developments in the South African mining industry' (University of Cape Town, 1987).
26. Philip A. Hirschson, 'Management ideology and environmental turbulence: Understanding labour policies in the South African gold mining industry' (Oxford University unpublished M.Sc. dissertation, 1988), p. 59.
27. McNamara, 'Inter-group violence among black employees on South African gold mines', p. 33.
28. McNamara, 'Inter-group violence among black employees on South African gold mines', p. 33.
29. AAC, 'Housing and accommodation practices on Anglo American gold mines in South Africa: Status report' (Johannesburg, 1988), p. 2.
30. AAC, 'Housing and accommodation practices', p. 1.
31. AAC, 'Housing and accommodation practices', p. 2.
32. Greenberg, Legitimating the Illegitimate, pp. 56–84; see also Douglas Hindson, Pass Laws and the African Proletariat (Johannesburg, 1987).
33. Jonathan Crush, 'Accommodating black miners: Home-ownership on the mines', in South African Review 5, eds. G. Moss and I. Obery (Johannesburg, 1989), pp. 335–8.
34. The account of mine accommodation in the 1980s is based on Crush, 'Accommodating black miners', pp. 335–47; Jonathan Crush and Wilmot G. James, 'Depopulating the compounds: Migrant labour and mine housing in South Africa', World Development, 19, no. 3 (1991), pp. 301–16; Wilmot G. James, 'Urban labour and the gold mining industry', Industrial Relations Journal of South Africa, 8, no. 4 (1988), pp. 15–30; Wilmot G. James, 'The future of migrant labour: Prospects and possibilities', Industrial Relations Journal of South Africa, 9, no. 4 (1989), pp. 15–26; William Cobbett, 'Trade unions, migrancy and the struggle for housing', in South African Review 5, pp. 322–34.
35. Lipton, Capitalism and Apartheid, p. 134–7; Crush, 'Migrancy and militance', p. 9.
36. Crush, 'Home-ownership on the mines', pp. 340–1; AAC: 'Monthly housing status report' (January 1991).
37. Crush and James, 'Depopulating the compounds', p. 23.

38. Fion de Vletter, 'Foreign labour on the South African gold mines: New insights into an old problem', *International Labour Review*, 126, no. 2 (1987), pp. 199–219.

39. James, 'Urban labour and gold mining industry', p. 27.

40. AAC: Market Research Africa, 'Accommodation study', Parts 1 and 2 (Johannesburg, 1986).

41. See Francis Wilson, 'Mineral wealth, rural poverty', in *Up Against the Fences: Poverty, Passes and Privilege in South Africa*, eds. Hermann Giliomee and Lawrence Schlemmer (Cape Town, 1985), pp. 57–64.

42. Cobbett, 'Trade unions, migrancy and the struggle for housing', p. 330.

43. Cobbett, 'Trade unions, migrancy and the struggle for housing', p. 323.

44. TEBA, 'TEBA strategic plan 1988–2000' (Johannesburg, 1988).

45. Centro de Estudos Africanos, 'The South African mining industry and Mozambican labour in the 1980s: An analysis of recent trends in employment policy' (Geneva, International Labour Organisation, 1987).

46. Tony Fleischer, 'TEBA: The Employment Bureau of Africa', *Mining Survey*, 77, no. 2 (1975), p. 22.

Chapter 6

1. Friedman, *Building Tomorrow Today*, p. 355.

2. Chamber of Mines, 'Memorandum to the Riekert Commission' (Johannesburg, 12 January 1978), p. 14.

3. See Transvaal Chamber of Mines, 'Tribal natives and trade unionism: The policy of the Rand gold mining industry' (Johannesburg, 1946), p. 1; 'The Native workers on the Witwatersrand gold mines' (Johannesburg, 1947); Gold Producers Committee, 'Statements of evidence: Native Laws Commission of Enquiry' (Johannesburg, 1947); *Annual Report* (Johannesburg, 1951).

4. Crush, 'Migrancy and militance: The case of the National Union of Mineworkers of South Africa', p. 8.

5. Roger Southall, 'Migrants and trade unions in South Africa today', *Canadian Journal of African Studies*, 20, no. 5 (1986), pp. 175–9.

6. Leger, 'Key issues in safety and health in South African mines', pp. 1–48.

7. Eddie Webster, 'The rise of social-movement unionism: The two faces of the black trade union movement in South Africa', in *State, Resistance and Change in South Africa*, eds. P. Frankel, N. Pines and M. Swilling (London, 1988), pp. 174–96.

8. See in particular John Saul and Stephen Gelb, *The Crisis in South Africa: Class Defense, Class Revolution* (New York and London, 1981), pp. 1–8.

9. McNamara, 'Inter-group violence among black employees on South African gold mines', pp. 23–38.

10. See Mike Kirkwood, 'The mine worker's struggle', *South African Labour Bulletin*, 1, no. 8 (1975), pp. 29–41.

11. Friedman, *Building Tomorrow Today*, p. 357.

12. For a discussion of state officials and their drive to privatise and deregulate industrial relations, see Greenberg, *Legitimating the Illegitimate*, pp. 13–18, 179–96.

13. AAC: D. Etheredge, Memorandum for D. B. Joffe, 22 June 1977. Interview with Jeff Lever and Wilmot G. James (April 1987).

14. AAC, Note for J. F. Drysdale/C. W. H. du Toit, from Z. de Beer (2 September 1977). Interview with Jeff Lever and Wilmot G. James (April 1987).

15. AAC, Note for Drysdale and Du Toit.

16. AAC, Note for D. A. Etheredge, from P. L. Nathan (10 August 1977).

17. AAC, Note for D. M. Rood, 'The role of AAC SA Ltd in the representations to the Wiehahn Commission on behalf of the mining industry', from K. H. Williams (4 July 1977).

18. AAC, Note for D. A. Etheredge, from Z. de Beer (2 September 1977).

19. Chamber of Mines, 'Statement of evidence to the Wiehahn Commission' (draft) (September 1978), Chapter VI, p. 17.

20. Chamber of Mines, 'Statement of evidence' (draft).

21. Chamber of Mines, 'Statement of evidence' (draft), Chapter VI, p. 15.

22. Chamber of Mines, 'Statement of evidence' (draft), Chapter VI, p. 15.

23. Chamber of Mines, 'Statement of evidence to the Commission of Inquiry into labour legislation' (Johannesburg, 1978), Chapter VI, p. 9.

24. Crush, 'Migrancy and militance', p. 6.

25. COMRO, *Factors Affecting the Popularity of a Mine for Black Mineworkers* (Johannesburg, Report 58/77, January 1978); COMRO, 'Factors which black workers viewed as contributing to and detracting most from the quality of life on a gold mine' (Johannesburg, HRL Monitoring Report, April 1980).

26. Crush, 'Migrancy and militance', p. 10.

27. McNamara, 'Inter-group violence among black employees', p. 31.

28. Crush, 'Migrancy and militance', p. 13.

29. Leger and Van Niekerk, 'Organizing on the mines: The NUM phenomenon', p. 3.

30. Eddie Webster, 'The independent black trade union movement in South Africa: A challenge to management', in *A Question of Survival*, eds. Michel Albeldas and Alan Fischer (Johannesburg, 1987), p. 24.

31. Leger, 'Key issues in safety and health in South African mines', pp. 1–48. The following discussion of safety and health relies on Jean Leger's invaluable work; Leger, 'Towards safer underground work' (Johannesburg: University of the Witwatersrand, 1985); Leger, 'Safety and the organisation of work in South African mines: A crisis of control', *International Labour Review*, 125, no. 5 (1986), pp. 591–603.

32. Friedman, *Building Tomorrow Today*, p. 369; Leger, 'Safety and the organisation of work in South African mines', pp. 595–600.

33. Golding, 'Mass dismissals on the mines', p. 97.

34. Nicholas Haysom, 'The Industrial Court: Institutionalising industrial conflict', in *South African Review Two*, ed. South African Research Service (Johannesburg, 1984), p. 113.

35. SAIRR, *Race Relations Survey 1985* (Johannesburg, 1986), p. 200; Paul Benjamin, 'Trade unions and the Industrial Court', *South African Review 4*, eds. G. Moss and I. Obery (Johannesburg, 1987), p. 259.

36. SAIRR, *Race Relations Survey 1986* (Johannesburg, 1987), p. 275.

37. SAIRR, *Race Relations Survey 1984*, p. 335.

38. SAIRR, *Race Relations Survey 1986*, p. 275.

39. Cyril Ramaphosa, 'The Industrial Court from the employees' point of view' (Johannesburg, Address to the Joint Bar Council, 1987).

40. Friedman, *Building Tomorrow Today*, p. 363.

41. Martin Nicol, 'Towards a wage policy in the mining industry', *South African Labour Bulletin*, 14, no. 3 (1989), pp. 67–75.

42. *Finance Week*, 29 April 1986.

43. Marcel Golding, quoted in *Business Day*, 12 June 1986.

44. Unpublished proceedings, Arbitration, National Union of Mineworkers and the

Anglo American Corporation (Johannesburg, 1988), Section PT202, p. 93.

45. The General Manager, Chamber of Mines of South Africa, to the General Secretary, NUM, 17 August 1987.

46. Gavin Relly, Annual General Meeting of the Anglo American Corporation (Johannesburg, 1987).

47. Quoted in *Financial Mail* (Supplement), 25 May 1986.

48. Unpublished proceedings, Arbitration, Section PT201, p. 61.

49. Chamber of Mines to Cyril Ramaphosa, 31 August 1987.

50. NUM, *Collective Bargaining at Anglo American Mines – A Model for Reform or Repression* (Johannesburg, 1988).

51. See also Karl van Holdt, 'Mine repression: NUM fights back', *South African Labour Bulletin*, 13, no. 8 (1989), pp. 81–91; and 'Interview: James Motlatsi, president of NUM', *South African Labour Bulletin*, 14, no. 2 (1989), pp. 22–35.

52. Nicol, 'Towards a wage policy in the mining industry', pp. 67–75.

53. Cyril Ramaphosa, in *A Question of Survival: Conversations with Key South Africans*, p. 291.

Chapter 7

1. McNamara, 'Inter-group violence among black employees on South African gold mines'. In managerial circles 'compounds' refer to the older barrack-style accommodation, while 'hostels' are used for the more modern, upgraded versions. However, since the newer versions are sociologically identical to the old, in the sense that their *social* functions have not changed, 'compounds' and 'hostels' will be used interchangeably in this narrative.

2. Wilson, *Labour in the South African Gold Mines*; Francis Wilson, *Migrant Labour* (Johannesburg, 1974); Johnstone, *Class, Race and Gold*; Webster, ed., *Essays in Southern African Labour History*; Lipton, 'Men of two worlds'.

3. Van Onselen, *Chibaro*, pp. 136–8.

4. Wilson, *Labour in the South African Gold Mines*, p. 57.

5. Johnstone, *Class, Race and Gold*, p. 169.

6. Moodie, 'The moral economy of the black miners' strike of 1946', pp. 29–30.

7. John Rex, 'The compound, reserve and urban location', *South African Labour Bulletin*, 1, no. 4 (1974); Johnstone, *Class, Race and Gold*; P. Pearson, 'Authority and control in a South African gold mine compound', in *Papers Presented at the African Studies Seminar* (Johannesburg, University of the Witwatersrand, 1977); Sean Moroney 'The compound as a mechanism of worker control', *South African Labour Bulletin*, 4, no. 3 (1978).

8. Rex, 'The compound, reserve and urban location', p. 8.

9. T. Dunbar Moodie, 'Mine culture and miners' identity on the South African gold mines', in *Town and Countryside in the Transvaal*, ed. B. Bozzoli (Johannesburg, 1983); Moodie, 'The moral economy of the black miners' strike'; R. Gordon, *Mines, Masters and Migrants: Life in a Namibian Compound* (Johannesburg, 1977); and Van Onselen, *Chibaro*.

10. Unpublished proceedings, Arbitration.

11. TEBA, 'SMR' (4th Quarter 1986).

12. Crush, 'Migrancy and militance: The case of the National Union of Mineworkers of South Africa', p. 10.

13. Descriptions of the earlier ethnic system of supervision can be found in Moodie,

'The formal and informal social structure of a South African gold mine'.

14. C. Ramaphosa, general secretary, National Union of Mineworkers, to General Manager, Chamber of Mines, 6 August 1987.
15. Unpublished proceedings, Arbitration, Section PT 202, p. 118.
16. For an account of the strike at a Gold Fields mine, which illustrates the power of managerial and security control at a mine where the NUM was weakly represented, see Mzimkulu Malunga, 'My life as a miner: Part IV', *Weekly Mail* (19–25 August 1988), pp. 10–11. See also unpublished proceedings, Arbitration, Section 202, p. 89. Gold Fields was also accused of tolerating the use of torture against miners. See *Weekly Mail* (13–19 February 1987), pp. 1–2, and the response of Gold Fields, *Business Day* (16 February 1987).
17. TEBA, 'SMR' (August–September 1987).
18. TEBA, 'SMR' (August 1987).
19. Unpublished proceedings, Arbitration, Section PT 203, p. 125.
20. Unpublished proceedings, Arbitration, Section PT 203, p. 140.
21. Management source cited in unpublished proceedings, Arbitration, Section PT 203.
22. Management source cited in unpublished proceedings, Arbitration, Section PT 204.
23. Union of South Africa, *Report of the Native Grievances Enquiry* (Pretoria, UG37/1914).
24. Union of South Africa, *Report of the Witwatersrand Mine Natives' Wages Commission* (Pretoria, UG22/1943).
25. AAC and NUM: Leatt *et al.*, 'Reaping the whirlwind? Report on a joint study by the National Union of Mineworkers and Anglo American Corporation.
26. Mine Management, Western Deep Levels, referred to in unpublished proceedings, Arbitration.
27. See Mancur Olson, *The Logic of Collective Behavior; Public Goods and the Theory of Groups* (Cambridge: Harvard University Press, 1971), pp. 70–1.
28. A term used C. Markham and M. Mothibeli, 'The 1987 mineworkers' strike', *South African Labour Bulletin*, 13, no. 1 (1987), pp. 58–75.
29. Unpublished proceedings, Arbitration, Section PT 203, pp. 124–5.
30. Unpublished proceedings, Arbitration, Section PT 203, p. 131.
31. See McNamara, 'Inter-group violence among black employees on South African gold mines', pp. 31–3.
32. The relationship between women and the mine hostel environment is an important but largely undeveloped theme. In the material consulted it was clear that women formed an essential part of the informal network established at mines. A majority of the women who visited men at the hostels were, in the material consulted, wives visiting husbands. For an account of male sexuality and women in the compounds, see Moodie, 'Migrancy and male sexuality on the South African gold mines'.
33. Mine management, Western Deep Levels, referred to in unpublished proceedings, Arbitration.
34. Unpublished proceedings, Arbitration, Section PT 201, p. 73.
35. Unpublished proceedings, Arbitration, Section PT 203, p. 125.
36. NUM, *Collective Bargaining at Anglo American Mines – A Model for Reform or Repression* (Johannesburg, 1988).

Chapter 8

1. I am indebted to Jeff Lever for much of the information and insights contained in this chapter. It is based in part on a monograph, Jeff Lever and Wilmot G. James,

'Towards a deracialised labour force: Industrial relations and the abolition of the colour bar in South Africa's gold industry' (University of Stellenbosch: Unit for the Sociology of Development, 1987).

2. Historical accounts of the colour bar can be found in Van der Horst, *Native Labour in South Africa*, pp. 173–85; W. H. Hutt, *The Economics of the Colour Bar* (London, 1964), pp. 58–65; G.V. Doxey, *The Industrial Colour Bar in South Africa* (Cape Town, 1961), pp. 110–47; Wilson, *Labour in the South African Gold Mines*, pp. 7–13, 110–19; and Johnstone, *Class, Race and Gold*, pp. 77–88, 93–166.

3. Lever, 'Established trade unions and industrial relations on the gold mines in the 1980s', p. 7.

4. Republic of South Africa, *Government Gazette* No. 11397 (8 July 1988), pp. 10–19; *Government Gazette*, No. 11504 (16 September 1988), pp. 12–16.

5. *Business Day*, 5 September 1989, p. 3.

6. Lipton, *Capitalism and Apartheid*, pp. 113–19.

7. See also Pycroft and Munslow, 'Black mine workers in South Africa: Strategies of co-option and resistance', pp. 159–60.

8. Wilson, *Labour in the South African Gold Mines*, pp. 115–19.

9. This narrative draws on Ari Sitas, 'Rebels without a pause: The white miner, changes in the labour process and conflict 1964–1978' (University of the Witwatersrand: unpublished B.A. Honours thesis, 1979).

10. Association of Mine Managers of the Transvaal, *Papers and Discussion 1963–64* (Johannesburg, 1964).

11. N. E. Wiehahn, *The Complete Wiehahn Report* (Pretoria, 1982) pp. 694–8. Data was supplied by the Chamber of Mines Research Organisation on the request of the Commission. Chamber of Mines, 'Oral evidence to the Wiehahn Commission' (23 February 1978).

12. Chamber of Mines, *Annual Report* (Johannesburg, 1968), p. 10.

13. Chamber of Mines, *Annual Report* (Johannesburg, 1970), p. 12.

14. Chamber of Mines, *Annual Report* (Johannesburg, 1968), p. 9; *Annual Report* (Johannesburg, 1969), p. 8; *Annual Report* (Johannesburg, 1970), p. 12; *Annual Report* (Johannesburg, 1971), p. 10.

15. Chamber of Mines, *Annual Report* (Johannesburg, 1970), p. 12.

16. Chamber of Mines, *Annual Report* (Johannesburg, 1970), p. 13; *Annual Report* (Johannesburg, 1968), p. 8.

17. Wilson, *Labour in the South African Gold Mines*, p. 119.

18. Wiehahn, *The Complete Wiehahn Report*, p. 709.

19. T. Mizrahan and P. de Vries, 'Report on the job of artisan aide in gold mining' (Johannesburg, Report 52/76, 1976).

20. Chamber of Mines, Members' circulars: Letter to the Federation of Mining Unions (25 June 1973).

21. Chamber of Mines, Members' circulars: Letter to the Federation of Mining Unions.

22. RSA, *Report and Recommendation by the Industrial Tribunal to the Honourable the Minister of Labour on Reservation of Work Connected to Sampling, Surveying and Ventilation on Mines, Republic of South Africa* (Pretoria, 1971), p. 40.

23. Chamber of Mines, Members' circulars: Letter from the General Manager to D. A. Etheredge (11 May 1985).

24. RSA, *Report of the Commission of Inquiry into the Possible Introduction of a Five-Day Working Week in the Mining Industry of the RSA* (Pretoria, RP97/1977), p. 2.

25. *Report of the Commission of Inquiry into the Possible Introduction of a Five-Day Working*

Week, pp. 137–9.

26. Wiehahn, *The Complete Wiehahn Report*, pp. 713, 720–1.

27. For a review of changes in the labour process, see Michael O'Donovan, 'The politics of the labour process in mining' (University of the Witwatersrand: unpublished B.A. Honours thesis, 1985).

28. Pundy Pillay, 'Future labour developments in the South African mining industry' (Cape Town, University of Cape Town, 1987), p. 4.

29. Moodie, 'The formal and informal social structure of a South African gold mine', p. 565.

30. O'Donovan, 'The politics of the labour process', p. 51.

31. AAC, 'Proposed approach to staffing mechanised development' (Johannesburg, 22 October 1975), p. 3.

32. AAC: 'Proposed approach to staffing mechanised development', p. 3.

33. D. L. van Coller, 'Job evaluation and the changing wage pattern', *Journal of the South African Institute of Mining and Metallurgy*, 75, no. 1 (1974), pp. 1–3.

34. Jacques Perold, 'The historical and contemporary use of job evaluation in South Africa', *South African Labour Bulletin*, 10, no. 4 (1985), pp. 72–92.

35. Chamber of Mines, *Annual Report* (Johannesburg, 1978), p. 31.

36. *Die Mynwerker*, 7 October 1981.

37. *Mine Surface Officials Association (MSOA) Journal*, May 1977.

38. *MSOA Journal*, April 1984.

39. Chamber of Mines, 'Preliminary notes on statements of evidence to the Wiehahn Commission' (Johannesburg, 9 February 1978), p. 13.

40. AAC, Memoranda: Note from P. L. Nathan to D. A. Etheredge (10 August 1977).

41. AAC, Memoranda: Note from Nathan to Etheredge.

42. Chamber of Mines, 'Members' Circular' (Johannesburg, 30 June 1977).

43. AAC, Memoranda: Memorandum for D. B. Hoffe from D. A. Etheredge (22 June 1977); Note from Nathan to Etheredge; Note from Z. de Beer to J. F. Drysdale and C. W. H. du Toit (2 September 1977); Chamber of Mines, 'Oral evidence to the Wiehahn Commission' (23 February 1978).

44. Wiehahn, *The Complete Wiehahn Report*, p. 724.

45. Wiehahn, *The Complete Wiehahn Report*, p. 760.

46. Craig Charney, 'The National Party', in *The State of Apartheid*, ed. Wilmot G. James (Boulder, 1987), p. 17.

47. *Steelworker*, October 1983; *Tecamp*, June 1984.

48. *New York Times*, 14 May 1985.

49. House of Assembly Debates (*Hansard*), Column 6335 (25 May 1985).

50. Johan Liebenberg, interviewed by Jeff Lever and Wilmot G. James (April 1987).

51. 'First Draft of Bill to Amend the Mines and Works Act', pp. 4–6.

52. Chamber of Mines, 'Evidence to the parliamentary standing committee on minerals and energy' (Cape Town, 22 August 1986).

53. Chamber of Mines, 'Core agreement' (Johannesburg, 25 July 1986).

54. Chamber of Mines, 'Draft Industrial Council security of employment agreement' (Johannesburg, 1986), pp. 11–14.

55. 'Draft Bill to Amend the Mines and Works Act' (Pretoria, 4 June 1986).

56. Cyril Ramaphosa and Marcel Golding, 'Evidence to the parliamentary standing committee on minerals and energy' (Cape Town, August 1987), p. 20.

57. Ramaphosa and Golding, 'Evidence to the PSCME', p. 1.

58. See Michael Burawoy, *The Colour of Class on the Copper Mines: From African Advan-*

cement to Zambianisation, Zambian Papers no. 7 (Manchester, 1972).

Chapter 9

1. NUM, 'Political policy' (Johannesburg, 1987), p. 1.
2. Robert Davies, 'Nationalisation, socialisation and the Freedom Charter', *South African Labour Bulletin*, 12, no. 2 (1987), pp. 85–106.
3. Davies, 'Nationalisation, socialisation and the Freedom Charter', p. 87.
4. See Burawoy, 'The hidden abode of underdevelopment: Labor process and the state in Zambia', pp. 142–4.
5. See also 'Cyril Ramaphosa on the NUM congress', *South African Labour Bulletin*, 12, no. 3 (1987), p. 54.
6. See Theda Skocpol, 'Bringing the state back in: Strategies in analysis and current research', in *Bringing the State Back In*, eds. Peter Evans, D. Rueschmeyer and T. Skocpol (Cambridge, 1985); Fred Block, 'Beyond relative autonomy: State managers as historical subjects', in *The Socialist Register*, eds. R. Miliband and J. Saville (London, 1980). A useful summary of this literature discussed in relation to the South African state can be found in Ivan Evans, 'Racial domination, capitalist hegemony and the state', *South African Sociological Review*, 1, no. 1 (1988), pp. 68–79.
7. See Georg Konrad and Ivan Szelyeni, *The Intellectuals on the Road to Class Power* (New York, 1979); Alvin Gouldner, *The Future of Intellectuals and the Rise of the New Class* (London, 1979).
8. See Michael Burawoy, 'Painting socialism in Hungary', *South African Labour Bulletin*, 15, no. 3 (1990), pp. 75–85; Michael Burawoy and János Lukács, 'The radiant future', *South African Sociological Review*, 3, no. 2 (1991), pp. 2–28.
9. Burawoy, 'The hidden abode of underdevelopment: Labor process and the state in Zambia', pp. 162–3.
10. NUM, 'Political policy'.
11. See for example, Barry Hindess, ed., *Reactions to the Right* (London, 1990).
12. Thabo Mbeki, as quoted in *Black Enterprise* (September 1990).
13. ANC, Department of Economic Policy, 'Discussion document on economic policy' (Harare, 1990), p. 9. For an analysis of state controls over marketing in Zimbabwe, see Peter Robins, 'The South African mining industry after apartheid', in *After Apartheid: Renewal of the South African Economy*, eds. Suckling and White, pp. 163–72.
14. Karl van Holdt, 'Insurrection, negotiation, and "war of position"', *South African Labour Bulletin*, 15, no. 3 (1990), p. 13. Van Holdt's argument is not new. Similar claims can be found in Heribert Adam, *Modernising Racial Domination: South Africa's Political Dynamics* (Berkeley, 1971); Heribert Adam and Kogila Moodley, *South Africa Without Apartheid: Dismantling Racial Domination* (Berkeley, 1986); Hermann Giliomee and Lawrence Schlemmer, eds., *Negotiating South Africa's Future* (Johannesburg, 1989).
15. Skocpol, *States and Social Revolutions*, especially pp. 47–154.
16. Roger J. Southall, 'Post-apartheid South Africa: Constraints on socialism', *Journal of Modern African Studies*, 25, no. 2 (1987), pp. 345–74.

Bibliography

A. Primary Sources

1 Official

Government Gazettes.

House of Assembly Debates (*Hansard*).

Evidence to the parliamentary standing committee on minerals and energy (August–September 1987).

Report of the Commission of Inquiry into Legislation Affecting the Utilization of Manpower (Riekert Commission) (Pretoria, RP32/1979).

Report of the Commission of Inquiry into Labour Legislation (Wiehahn Commission), Six Parts (Pretoria: RP47/1979, 38/1980, 32/1980, 87/1980, 27/1981, 28/1981).

Report and Recommendation by the Industrial Tribunal to the Honourable the Minister of Labour on Reservation of Work connected to Sampling, Surveying and Ventilation on Mines, Republic of South Africa (Pretoria, 1971).

Report of the Commission of Inquiry into the Possible Introduction of a Five-Day Working Week in the Mining Industry of the RSA (Franszen Commission), (Pretoria: Government Printer, RP97/1977).

Report of the Witwatersrand Mine Natives' Wages Commission (Pretoria: UG22/1943).

Report of the Native Grievances Enquiry (Pretoria: UG37/1914).

Commission of Enquiry into Economic Development of the Republic of Ciskei (Republic of Ciskei, 1983).

2 Chamber of Mines

Annual Report (series).

Members' circulars.

Statements of Evidence, Statistics and Memoranda Submitted to the Witwatersrand Mine Wages Commission (1943).

'Statement of evidence to Native Laws Commission' (1947).

'Memorandum to the Riekert Commission' (1978).

'Statements of evidence to the Wiehahn Commission' (1977–1980).

'Transcript of oral evidence to the Wiehahn Commission' (1979).

'Number of vacancies on gold mines: Artisans and officials' (Johannesburg, 1986).

'Evidence to the parliamentary standing committee on minerals and energy' (Cape Town, 22 August 1986).

'Draft Industrial Council security of employment agreement' (Johannesburg, 1986).

3 Chamber of Mines Research Organisation (COMRO)

Research and Development Annual Report (series)

Memoranda.

D. MacArthur, J. B. Godfrey, K. McNamara and O. F. Thomas, *Factors Affecting the Popularity of a Mine for Black Mineworkers*, Report No. 58/77 (January 1978).

W. Z. Mashaba and M. H. Steen, *ERPM Local Recruitment Experiment* (Human Research Laboratories [HRL] Project No. GH1A03, February 1986).

W. Z. Mashaba and M. H. Steen, *A Preliminary Investigation into the Feasibility of Attracting Urbanized Labour to Work in Mining* (HRL Project No. GH1A03, December 1986)

J. K. McNamara, 'Implications of the Rhodesian labour experiment for the recruitment of South African urban township residents into the gold mines' (COMRO: HRL, December 1985).

Factors Affecting the Popularity of a Mine for Black Mineworkers (Research Report No. 58/77, January 1978).

'Factors which black workers viewed as contributing most to and detracting most from the quality of life on a gold mine' (HRL Monitoring Report, April 1980).

T. Mizrahan and P. de Vries, *Report on the Job of Artisan Aide in Gold Mining* (Report No. 52/76).

4 The Employment Bureau of Africa (TEBA)

Annual Report and Financial Statement (series).

'Strategic Plan' (series) (1977–1988).

Memoranda.

'Summary of monthly field reports' ('SMR') (1985–1988).

Selected correspondence.

'Black labour force: Need for the industry to maintain the right to select its own labour force' (Johannesburg, 1984).

5 Anglo American Corporation (AAC)

Monthly housing status reports.

Commissioned accommodation reports.

Selected memoranda and notes.

6 National Union of Mineworkers (NUM)

Resolutions of Annual Congress (1985–1988).

Recognition agreements with Chamber of Mines.

Unpublished proceedings, Arbitration, National Union of Mineworkers and Anglo American Corporation (Johannesburg: 14–15 March 1988).

James Leatt, Paulus Zulu, Monoko Nchwe, Mark Ntshangase, and Richard Laughlin, 'Reaping the whirlwind? Report on a joint study by the National Union of Mineworkers and the Anglo American gold division on the causes of mine violence' (Johannesburg: NUM and AAC, 1986).

'Political Policy' (Johannesburg, 1987).

Selected correspondence with Chamber of Mines (1987).

Collective Bargaining at Anglo American Mines – A Model for Reform or Repression (9 December 1988).

7 African National Congress (ANC)
Department of Economic Policy, 'Discussion document on economic policy' (Harare, 1990).

8 Newspapers and Periodicals
Sunday Tribune.
Star.
Cape Times.
Business Day.
Black Enterprise.
New York Times.
Weekly Mail.
Finance Week.
Financial Mail.
Optima (Anglo American Corporation).
NUM News (National Union of Mineworkers).
Die Mynwerker (Mine Workers' Union).
MSOA Journal (Mine Surface Officials' Association).
Steelworker.
Tecamp (South African Technical Officials' Association).

B. Secondary Sources

Abrams, Philip, *Historical Sociology* (Ithaca, 1982).

Adam, Heribert, *Modernising Racial Domination: South Africa's Political Dynamics* (Berkeley, 1971).

Adam, Heribert, and Kogila Moodley, *South Africa Without Apartheid: Dismantling Racial Domination* (Berkeley, 1986).

Albeldas, Michael, and Alan Fischer, eds. *A Question of Survival: Conversations with Key South Africans* (Johannesburg, 1987).

Block, Fred, 'Beyond relative autonomy: State managers as historical subjects', in *The Socialist Register*, eds. R. Miliband and J. Saville (London, 1980).

Bohning, W. R., ed., *Black Migration to South Africa: A Selection of Policy-oriented Research* (Geneva, 1981).

Booth, Alan, 'Capitalism and the competition for Swazi labour, 1945–1960', *Journal of Southern African Studies*, 13, no. 1 (October 1986).

Bozzoli, Belinda, 'Marxism, feminism and South African studies', *Journal of Southern African Studies*, 9, no. 2 (1983).

Bradford, Helen, *A Taste of Freedom: The ICU in Rural South Africa, 1924–1930* (New Haven and Johannesburg, 1984).

Brochmann, Grete, 'Migrant labour and foreign policy: The case of Mozambique', *Journal of Peace Research*, 22, no. 4 (1985).

Brown, David, 'The basement of Babylon: Language and literacy on the South African gold mines', *Social Dynamics*, 14, no. 1 (June 1988).

Bundy, Colin, *The Rise and Fall of the South African Peasantry* (London, 1979).

Burawoy, Michael and János Lukács, 'The radiant future', *South African Sociological Review*, 3, no. 2 (1991).

Burawoy, Michael, 'Painting socialism in Hungary', *South African Labour Bulletin*, 15, no. 3 (1990).

Burawoy, Michael, *The Politics of Production: Factory Regimes under Capitalism and Socialism* (London, 1985).

Burawoy, Michael, 'The hidden abode of underdevelopment: Labor process and the state in Zambia', *Politics and Society*, 11, no. 2 (1982).

Burawoy, Michael, 'The functions and reproduction of migrant labor: Comparative material from southern Africa and the United States', *American Journal of Sociology*, 81, no. 5 (1977).

Burawoy, Michael, *The Colour of Class on the Copper Mines: From African Advancement to Zambianization*, Zambian Papers no. 7 (Manchester, 1972).

Centro de Estudos Africanos, 'The South African mining industry and Mozambican labour in the 1980s: An analysis of recent trends in employment policy' (Geneva, 1987).

Charney, Craig, 'The National Party', in *The State of Apartheid*, ed. Wilmot G. James (Boulder, 1987).

Chirimuuta, Richard, *Aids, Africa and Racism* (Derbyshire, 1987).

Christiaansen, Robert E., and Jonathan G. Kydd, 'The return of Malawian labour from South Africa and Zimbabwe', *Journal of Modern African Studies*, 21, no. 2 (1983).

Cobbe, James, 'Consequences for Lesotho of changing South African labour demand', *African Affairs*, 85, no. 338 (1986).

Cobbett, William, 'Trade unions, migrancy and the struggle for housing', in *South African Review 5*, eds. G. Moss and I. Obery (Johannesburg, 1989).

Cohen, R., *The New Helots: Migrants in the International Division of Labour* (Aldershot, 1988).

Crush, Jonathan, Alan Jeeves and David Yudelman, *South Africa's Labor Empire: A History of Black Mineworkers under Apartheid* (Boulder, forthcoming).

Crush, Jonathan, and Wilmot G. James, 'Depopulating the compounds: Migrant labour and mine housing in contemporary South Africa', *World Development*, 19, no. 3 (1991).

Crush, Jonathan, 'Accommodating black miners: Home-ownership on the mines', in *South African Review 5*, eds. G. Moss and I. Obery (Johannesburg, 1989).

Crush, Jonathan, 'Migrancy and militance: The case of the National Union of Mineworkers of South Africa', *African Affairs*, 88, no. 350 (1989).

Crush, Jonathan, 'Restructuring migrant labour on the gold mines', in *South African Review 4*, eds. G. Moss and I. Obery (Johannesburg, 1987).

Crush, Jonathan, 'The extrusion of foreign labour from the South African gold mining industry', *Geoforum*, 17, no. 2 (1986).

Crush, Jonathan, 'Swazi migrant workers and the Witwatersrand gold mines 1886–1920', *Journal of Historical Geography*, 12 (1986).

Davies, Robert, 'Nationalisation, socialisation and the Freedom Charter', *South African Labour Bulletin*, 12, no. 2 (1987).

Davies, Robert, 'South African strategy towards Mozambique since Nkomati', *Transformation*, 3 (1987).

Davies, Robert, *Capital, State and White Labour in South Africa 1900–1960* (Brighton, 1979).

Davies, Robert, David Kaplan, Mike Morris, and Dan O'Meara, 'Class struggle and the periodisation of the South African state', *Review of African Political Economy*, 7 (1978).

De Vletter, Fion, 'Foreign labour on the South African gold mines: New insights into an old problem', *International Labour Review*, 126, no. 2 (1987).

De Vletter, Fion, 'Recent trends and prospects of black migration to South Africa', *Journal of Modern African Studies*, 23 (1985).

Doxey, G. V., *The Industrial Colour Bar in South Africa* (Cape Town, 1961).

Evans, Ivan, 'Racial domination, capitalist hegemony and the state', *South African Sociological Review*, 1, no. 1 (1988).

First, Ruth, *Black Gold: The Mozambican Miner, Peasant or Proletarian* (Brighton, 1983).

Fleischer, Tony, 'TEBA – The Employment Bureau of Africa', *Mining Survey*, 77, no. 2 (1975).

Friedman, Steven, *Building Tomorrow Today: African Workers in Trade Unions 1970–1984* (Johannesburg, 1987).

Friedman, Steven, 'Chamber of Mines' policy and the emerging miners' unions', *South African Labour Bulletin*, 8, no. 5 (April 1983).

Giliomee, Hermann, and Lawrence Schlemmer, eds., *Negotiating South Africa's Future* (Johannesburg, 1989).

Golding, Marcel, 'Mass dismissals on the mines', *South African Labour Bulletin*, 10, no. 7 (1985).

Gordon, R., *Mines, Masters and Migrants: Life in a Namibian Compound* (Johannesburg, 1977).

Gouldner, Alvin, *The Future of Intellectuals and the Rise of the New Class* (London, 1979).

Green, Timothy, *The World of Gold* (London, 1980).

Greenberg, Stanley B., *Legitimating the Illegitimate: State, Markets and Resistance in South Africa* (Berkeley, 1987).

Greenberg, Stanley B., *Race and State in Capitalist Development: Comparative Perspectives* (New Haven, 1980).

Greenberg, Stanley, B., and Hermann Giliomee, 'Managing influx control from the rural end: The black homelands and the underbelly of privilege', in *Up Against the Fences: Poverty, Passes and Privilege in South Africa*, eds. H. Giliomee and L. Schlemmer (Cape Town, 1985).

Guy, Jeff, and Motlatsi Thabane, 'Technology, ethnicity and ideology: Basotho miners and shaft-sinking on the South African gold mines', *Journal of Southern African Studies*, 14, no. 2 (1988).

Haysom, Nicholas, 'The Industrial Court: Institutionalising industrial conflict', in *South African Review Two*, ed. South African Research Service (Johannesburg, 1984).

Hindess, Barry, ed. *Reactions to the Right* (London, 1990).

Hindson, Douglas, *Pass Laws and the African Proletariat* (Johannesburg, 1987).

Hutt, W. H., *The Economics of the Colour Bar* (London, 1964).

Hyslop, J., 'A Prussian path to apartheid? Germany as comparative perspective in critical analysis of South African society', *South African Sociological Review*, 3, no. 1 (1990).

Innes, Duncan, *Anglo American and the Rise of Modern South Africa* (Johannesburg, 1984).

Interview with Cyril Ramaphosa, *South African Labour Bulletin*, 12, no. 3 (1987).

James, Wilmot G., 'The future of migrant labour: Prospects and possibilities', *Industrial Relations Journal of South Africa*, 9, no. 4 (1989).

James, Wilmot G., 'Urban labour and the gold mining industry', *Industrial Relations Journal of South Africa*, 8, no. 4 (1988).

James, Wilmot G., 'Grounds for a strike: South African gold mining in the 1940s', *African Economic History*, no. 17 (1987).

Jeeves, Alan, *Migrant Labor in South Africa's Mining Economy* (Kingston and Montreal, 1985).

Johnson, Bekki, *AIDS in Africa: A Review of Medical, Public Health, Social Science and Popular Literature* (Aachen, 1988).

Johnstone, Frederick A., *Class, Race and Gold: A Study of Class Relations and Racial Discrimination in South Africa* (London, 1976).

Johnstone, Frederick A., 'White prosperity and white supremacy in South Africa today', *African Affairs*, 69 (1970).

Kirkwood, Mike, 'The mine workers' struggle', *South African Labour Bulletin*, 1, no. 8 (1975).

Koch-Weser, Dieter, and Hennelore Vanderschmidt, eds., *The Heterosexual Transmission of AIDS in Africa* (Cambridge, 1988).

Konrad, Georg and Ivan Szelyeni, *The Intellectuals on the Road to Class Power* (New York, 1979).

Lambert, Rob, 'Trade unions, nationalism, and the socialist project in South Africa', *South African Review 4*, eds. G. Moss and I. Obery (Johannesburg, 1987).

Legassick, Martin, 'South Africa: Forced labour, industrialization and racial differentiation', in *The Political Economy of Africa*, ed. R. Harris (Boston, 1973).

Leger, Jean, 'Key issues in safety and health in South African mines', *South African Sociological Review*, 2, no. 2 (1990).

Leger, Jean, 'Mozambican miners' reprieve', *South African Labour Bulletin*, 12, no. 2 (January–February 1987).

Leger, Jean, *Towards Safer Underground Work* (University of the Witwatersrand, 1985).

Leger, Jean, 'Safety and the organisation of work in South African gold mines: A crisis of control', *International Labour Review*, 125, no. 5 (September–October 1986).

Leger, Jean, and Philip van Niekerk, 'Organising on the mines: The NUM phenomenon', in *South African Review 3* (Johannesburg, 1985).

Lever, Jeffrey, 'Established trade unions and industrial relations on the gold mines in the 1980s', *Industrial Relations Journal of South Africa*, 8, no. 2 (1988).

Lever, Jeffrey, 'Artisan unions since Wiehahn', *Industrial Relations Journal of South Africa*, 1st Quarter (1985).

Lever, Jeffrey, and Wilmot G. James, 'Towards a deracialised labour force: Industrial relations and the abolition of the job colour bar on the South African gold mines', Occasional paper no. 12 (Stellenbosch, 1987).

Levy, Norman, *The Foundations of the South African Cheap Labour System* (London, 1982).

Lewis, Jon, 'South African labour history: An historiographical assessment', *Radical History Review*, 46/7 (Winter 1990).

Lewis, Jon, *Industrialization and Trade Union Organization in South Africa, 1924–1955: The Rise and Fall of the South African Trades and Labour Council* (Cambridge, 1984).

Lipton, Merle, *Capitalism and Apartheid: South Africa 1910–1986* (Aldershot and Cape Town, 1985 and 1989).

Lipton, Merle, 'Men of two worlds: Migrant labour in South Africa', *Optima*, 29, nos. 2 and 3 (1980).

Lodge, Tom, 'Political mobilisation in the 1950s: An East London case study', in *The Politics of Race, Class and Nationalism in Twentieth Century South Africa*, eds. S. Marks and S. Trapido (London, 1987).

Makanjee, V., 'Bophuthatswana: Bordering on no-man's land', *Indicator South Africa*, 5, no. 4 (1988).

Mann, Michael, *States, War and Capitalism: Studies in Political Sociology* (Oxford, 1988).

Maree, Johann, ed. *The Independent Trade Union Movement: Ten Years of the Labour Bulletin* (Johannesburg, 1989).

Markham, Coletane, and Monyaola Mothibeli, 'The 1987 mineworkers' strike', *South African Labour Bulletin*, 13, no. 1 (1987).

Martiny, J., and J. Sharp, 'An overview of QwaQwa: Town and country in a South African bantustan' (Cape Town, Carnegie Conference Paper no. 286, 1984).

McNamara, J. K., 'Inter-group violence among black employees on South African gold mines 1974–1986', *South African Sociological Review*, 1, no. 1 (October 1988).

Miller, Norman, and Richard Rockwell, eds., *AIDS in Africa: The Social and Policy Impact* (Lewiston, NY, 1988).

Moodie, Dunbar, 'Migrancy and male sexuality on the South African gold mines', *Journal of Southern African Studies*, 14, no. 2 (1988).

Moodie, Dunbar, 'The moral economy of the black miners' strike of 1946', *Journal of Southern African Studies*, 13, no. 1 (October 1986).

Moodie, Dunbar, 'Mine culture and miners' identity on the South African gold mines', in *Town and Countryside in the Transvaal*, ed. Belinda Bozzoli (Johannesburg, 1983).

Moodie, Dunbar, 'The formal and informal social structure of a South African gold mine', *Human Relations*, 33, no. 8 (1980).

Moroney, Sean, 'The compound as a mechanism of worker control', *South African Labour Bulletin*, 4, no. 3 (1978).

Murray, Colin, 'Migrant labour and changing family in the rural periphery of southern Africa', *Journal of Southern African Studies*, 6, no. 2 (April 1980).

Murray, Colin, 'Migration, differentiation and the development cycle in Lesotho', *African Perspectives*, 2 (1978).

Nattrass, Jill, and Julian May, 'Migration and dependency: Sources and levels of income in Kwazulu', *Development Southern Africa*, 3, no. 4 (1986).

Nichol, Martin, 'Towards a wage policy in the mining industry', *South African Labour Bulletin*, 14, no. 3 (1989).

O'Meara, Dan, 'The 1946 African mineworkers' strike and the political economy of South Africa', *Journal of Commonwealth and Comparative Politics*, 13 no. 2 (1975).

Olson, Mancur, *The Logic of Collective Behavior: Public Goods and the Theory of Groups* (Cambridge, 1971).

Packard, Randall, *White Plague, Black Labor: Tuberculosis and the Political Economy of Health and Disease in South Africa* (Pietermaritzburg, Berkeley and London, 1990).

Packard, Randall, 'Tuberculosis and the development of industrial health policies on the Witwatersrand, 1902–1932', *Journal of Southern African Studies*, 13, no. 2 (1987).

Parson, Jack, 'The peasantariat and politics: Migration, wage labour and agriculture in Botswana', *Africa Today*, 31, no. 4 (1984).

Parpart, J. L., *Class Consciousness among Zambian Copper Miners 1950–1966*, Zambian working papers no. 53 (Lusaka, 1982).

Pearson, P., 'Authority and control in a South African goldmine compound', in *Papers Presented at the African Studies Seminar* (Johannesburg, 1977).

Perold, Jacques, 'The historical and contemporary use of job evaluation in South Africa', *South African Labour Bulletin*, 10, no. 4 (1985).

Pillay, Pundy, 'Future labour developments in the South African mining industry' (University of Cape Town, 1987).

Platzky, Laureen, and Cherryl Walker, *The Surplus People: Forced Removals in South Africa* (Johannesburg, 1985).

Pycroft, C. and B. Munslow, 'Black mine workers in South Africa: Strategies of co-option and resistance', *Journal of Asian and African Studies*, 23, nos. 1 and 2 (1988).

Rafel, R., 'Job reservation on the mines', in *South African Review 4*, eds. G. Moss and I. Obery (Johannesburg, 1987).

Ragin, Charles, *The Comparative Method: Moving Beyond Qualitative and Quantitative*

Strategies (Berkeley, 1987).

Ranger, Terence and Eric Hobsbawn, *The Invention of Tradition* (New York, 1983).

Rex, John, 'The compound, reserve and urban location', *South African Labour Bulletin*, 1, no. 4 (1974).

Robins, Peter, 'The South African mining industry after apartheid', in *After Apartheid: Renewal of the South African Economy*, eds. John Suckling and Landeg White (London, 1988).

Roux, Eddie, *Time Longer than Rope* (Madison, 1964).

Saul, John, and Stephen Gelb, *The Crisis in South Africa: Class Defense, Class Revolution* (New York and London, 1981).

Sharp, J., 'Relocation and the problem of survival in QwaQwa: A report from the field', *Social Dynamics*, 8, no. 2 (1982).

Sharp, J. and A. Spiegel, 'Vulnerability to impoverishment in South African rural areas' (Cape Town, Carnegie conference paper no. 52, 1984).

Simkins, Charles, *Four Essays on the Past, Present and Future Distribution of the African Population of South Africa* (Cape Town, 1983).

Simkins, Charles, 'What has been happening to income distribution and poverty in the homelands?' (Cape Town, Carnegie conference paper no. 7, 1984).

Simkins, Charles, 'Agricultural production in the African reserves of South Africa', *Journal of Southern African Studies*, 7, no. 2 (1981).

Simons, H. J. and R. E., *Class and Colour in South Africa 1850–1950* (Harmondsworth, 1968).

Sitas, Ari, 'Moral formations and struggles amongst migrant workers on the East Rand', *Labour, Capital and Society*, 18, no. 2 (1985).

Skocpol, Theda, 'Bringing the state back in: Strategies in analysis and current research', in *Bringing the State Back In*, eds. Peter Evans, D. Rueschmeyer and T. Skocpol (Cambridge, 1985).

Skocpol, Theda, ed., *Vision and Method in Historical Sociology* (New York, 1984).

Skocpol, Theda, *States and Social Revolutions* (Cambridge, 1980).

South African Institute of Race Relations (SAIRR), *Race Relations Survey 1985* (Johannesburg, 1986).

South African Institute of Race Relations (SAIRR), *Race Relations Survey 1986* (Johannesburg, 1987).

South African Institute of Race Relations (SAIRR), *Race Relations Survey 1987/1988* (Johannesburg, 1988).

South African Institute of Race Relations (SAIRR), *Race Relations Survey 1988/1989* (Johannesburg, 1989).

South African Research Service (SARS), *South African Review 1* (Johannesburg, 1983).

South African Research Service (SARS), *South African Review 2* (Johannesburg, 1984).

South African Research Service (SARS), *South African Review 3* (Johannesburg, 1985).

Southall, Roger J., 'Post-apartheid South Africa: Constraints on socialism', *Journal of Modern African Studies*, 25, no. 2 (1987).

Southall, Roger J., 'Migrants and trade unions in South Africa today', *Canadian Journal of African Studies*, 20, no. 5 (1986).

Suckling, John and Landeg White, eds., *After Apartheid: Renewal of the South African Economy* (London, 1988).

Technical Assistance Group (TAG), *New Technologies on the Mines* (Johannesburg, 1986).

Thompson, Clive, 'Black trade unions on the mines', *South African Review 2*, ed. SARS (Johannesburg, 1984).

Trapido, S., 'South Africa in a comparative study of industrialisation', *Journal of Development Studies*, 7, no. 3 (1971).

Turrell, Robert, *Capital and Labour on the Kimberley Diamond Fields 1871-1890* (Cambridge, 1987).

Van Coller, D.L., 'Job evaluation and the changing wage pattern', *Journal of the South African Institute of Mining and Metallurgy*, 75, no. 1 (1974).

Van der Horst, Sheila, *Native Labour in South Africa* (Cape Town, 1942).

Van Onselen, Charles, *Chibaro: African Mine Labour in Southern Rhodesia 1900-1933* (London, 1976).

Van Onselen, Charles, *Studies in the Social and Economic History of the Witwatersrand: New Nineveh, New Babylon* (London, 1982).

Van Holdt, Karl, 'Insurrection, negotiation, and "war of position"', *South African Labour Bulletin*, 15, no. 3 (1990).

Van Holdt, Karl, 'Mines' repression: NUM fights back', *South African Labour Bulletin*, 13, no. 8 (1989).

Webster, Eddie, 'The rise of social-movement unionism: The two faces of the black trade union movement in South Africa', in *State, Resistance and Change in South Africa*, eds. P. Frankel, N. Pines and M. Swilling (London, 1988).

Webster, Eddie, 'The independent black trade union movement in South Africa: A challenge to management', in *A Question of Survival: Conversations with Key South Africans*, eds. Michel Albeldas and Alan Fischer (Johannesburg, 1987).

Webster, Eddie, *Cast in a Racial Mould: Labour Process and Trade Unionism in the Foundries* (Johannesburg, 1985).

Webster, Eddie, ed., *Essays in Southern African Labour History* (Johannesburg, 1978).

Whiteside, A. and C. Patel, 'Agreements concerning the employment of foreign black labour in South Africa' (Geneva, 1985).

Wiehahn, N. E., *The Complete Wiehahn Report* (Pretoria, 1982).

Wilson, Francis, and Mamphela Ramphele, *Uprooting Poverty: The South African Challenge* (Cape Town, New York, and London, 1989).

Wilson, Francis, 'Mineral wealth, rural poverty', in *Up Against the Fences: Poverty, Passes and Privilege in South Africa*, eds. H. Giliomee and L. Schlemmer (Cape Town, 1985).

Wilson, Francis, *Migrant Labour* (Johannesburg, 1974).

Wilson, Francis, *Labour in the South African Gold Mines 1911-1969* (Cambridge, 1972).

Wolpe, Harold, 'Capitalism and cheap labour-power in South Africa: From segregation to apartheid', *Economy and Society*, 1, no. 4 (November 1972).

Worger, William, *South Africa's City of Diamonds: Mine Workers and Monopoly Capitalism in Kimberley 1867-1895* (New Haven, 1987).

Yudelman, David, 'State and capital in modern South Africa', in *Democratic Liberalism in South Africa*, eds. Jeffrey Butler, Richard Elphick and David Welsh (Middletown, 1987).

Yudelman, David, and Alan Jeeves, 'New labour frontiers for old: Black migrants to the South African mines, 1920-1985', *Journal of Southern African Studies*, 13, no. 1 (October 1986).

Yudelman, David, *The Emergence of Modern South Africa: State, Capital and the Incorporation of Organized Labor on the South African Gold Fields, 1902-1939* (Westport and Cape Town, 1983).

C. Unpublished Works

Frost, David, 'The political economy of trackless mining (TM3) in the South African gold mines' (Cape Town: unpublished paper, 1988).

Frost, David, 'Hesistant revolution: Mechanisation, labour and the gold mines' (University of Cape Town: unpublished B.A. Honours thesis, 1987).

Hirschson, Philip A., 'Management ideology and environmental turbulence: Understanding labour policies in the South African gold mining industry' (Oxford University: unpublished M.Sc. thesis, 1988).

James, Wilmot G., 'Migrant labour selection in South Africa's metal mining industry' (Kingston, Canadian Association of African Studies, 1988).

James, Wilmot G., 'Politics and economics of internalisation: Labour migrancy to the South African gold mines 1980–2000' (Bellville, Association for Sociology in South Africa, 1987).

Mariotti, Amelia, 'The incorporation of African women into wage employment in South Africa, 1920–1970' (University of Connecticut: unpublished Ph.D. dissertation, 1980).

May, J. D., 'Migrant labour in Transkei – Causes and consequences on a village level' (University of Natal, 1985).

McNamara, J. K., 'Black worker conflicts on South African gold mines' (University of the Witwatersrand: unpublished Ph.D. dissertation, 1985).

O'Donovan, Michael, 'The labour process in gold mining' (University of the Witwatersrand: unpublished B.A. Honours dissertation, 1985).

Sitas, Ari, 'Rebels without a pause: The white miner, changes in the labour process and conflict 1964–1978' (University of the Witwatersrand: unpublished B.A. Honours thesis, 1979).

Index